FRETTING FATIGUE

FRETTING FATIGUE

Edited by
R. B. WATERHOUSE
Department of Metallurgy and Materials Science,
University of Nottingham, UK

APPLIED SCIENCE PUBLISHERS LTD
LONDON

APPLIED SCIENCE PUBLISHERS LTD
RIPPLE ROAD, BARKING, ESSEX, ENGLAND

British Library Cataloguing in Publication Data

Fretting fatigue.
1. Materials—Fatigue
I. Waterhouse, Robert Barry
620.1′123 TA418.38

ISBN 0-85334-932-0

WITH 12 TABLES AND 171 ILLUSTRATIONS

Printed in Great Britain by Galliard (Printers) Ltd, Great Yarmouth

Preface

When I first became interested in fretting corrosion, over a quarter of a century ago, most of the reported examples were concerned with the production of debris and the wear suffered by the mating surfaces. Experimental investigators used the quantitative measurement of the amount of wear as a means of assessing the severity of damage in a particular combination of materials. The influence of fretting on fatigue was scarcely touched upon, apart from the preliminary work of Warlow-Davies in 1938. He found that steel specimens subjected to fretting damage and subsequently tested in fatigue showed a reduction in fatigue strength of some 15 %. It soon became apparent that simultaneous fretting with fatigue could result in much greater reductions in fatigue strength. In this respect fretting fatigue resembled corrosion fatigue, i.e. there was a synergistic effect, or, as a corrosion scientist would say, it was a case of conjoint action. One of the important observations of the post-war period was that assessment of susceptibility to fretting fatigue was not related to the amount of fretting wear. In fact production of wear debris often resulted in little or no diminution in fatigue strength.

The possibility of fretting occurring in structures and multi-component systems which are subjected to vibration is so likely that the fatigue behaviour of such assemblages can be said to be entirely determined by the response of the contacting surfaces to fretting. Twenty-five years ago engineers and designers were not generally aware of this limitation, but some notable fatigue failures in the ensuing years have alerted them to this possibility, and it is now general practice to take into account the possibility of fretting at the design stage.

The purpose of the present volume is to give the most recent ideas on forecasting fretting fatigue damage, recognising it when it does occur and indicating some of its characteristics so that appropriate measures can be taken to avoid it.

I would like to express my appreciation to the contributors and also to Mrs J. Wheatley, Mrs J. A. Thornhill, Dr M. M. Hamdy and Dr D. E. Taylor for the help they have given in preparing the manuscript.

R. B. WATERHOUSE

Contents

List of Contributors

P. R. EDWARDS

Principal Scientific Officer, Structures Department, Royal Aircraft Establishment, Farnborough, Hants. GU14 6TD, UK.

K. ENDO

Professor, Department of Mechanical Engineering, Kyoto University, Kyoto, Japan.

P. J. E. FORSYTH

Procurement Executive, Materials Department, Royal Aircraft Establishment, Farnborough, Hants. GU14 6TD, UK.

D. W. HOEPPNER

Cockburn Professor of Engineering Design, Department of Mechanical Engineering, University of Toronto, Toronto, Ontario M5S 1A4, Canada.

J. J. O'CONNOR

Lecturer in Engineering Science, Department of Engineering Science, Oxford University, Parks Road, Oxford OX1 3PJ, UK.

D. E. TAYLOR

Research Fellow, The Wolfson Institute of Interfacial Technology, University of Nottingham, University Park, Nottingham NG7 2RD, UK.

R. B. Waterhouse

Reader in Metallurgy, Department of Metallurgy and Materials Science, University of Nottingham, University Park, Nottingham NG7 2RD, UK.

1

Introduction to Fatigue

R. B. WATERHOUSE

Department of Metallurgy and Materials Science,
University of Nottingham, UK

Of all the modes of mechanical failure, fatigue is the one most well known to the public mind. After any disaster such as a plane crash or bridge collapse the phrase most frequently seen in the newspapers is 'metal fatigue'. Nevil Shute's novel *No Highway*, in which Mr Honey predicted the imminent failure of the tail section of a plane in flight across the Atlantic, accustomed everyone to the idea that such normally unpredictable (except, of course, to Mr Honey) failures could occur with usually catastrophic consequences. Almost uncannily there followed the Comet disasters of the early 1950s. Despite the progress that has been made in the last 30 years in understanding the basic processes involved in fatigue and the development of fracture mechanics which enables one to say whether a flaw or crack will progress to a propagating crack, no one could yet claim to equal the precision of Mr Honey in his predictive powers. Even when the design is impeccable and the material, like Caesar's wife, above suspicion, there are still factors in service which render prediction difficult, e.g. random loading, corrosion and, of course, fretting.

There are two aspects of fatigue from the technical point of view. First, the provision of design data to enable engineers to construct machines and devices with the maximum degree of safety and economic life expectancy, and secondly the investigation of the basic mechanism of fatigue failure to allow the possible development of materials resistant to this type of failure.

1.1 DESIGN DATA

1.1.1 Stress Concentration

Initially the design data consisted of a knowledge of the stress concentration produced by holes and machined notches in a suitable test-piece. The fatigue strength of a particular material is generally determined from the S–log N curve obtained by testing suitable specimens in rotating–bending, plane-bending or push–pull machines at frequencies of 25–50 Hz, such methods and frequencies bearing some relation to the type of loading a machine element might experience in practice. The tests are carried out on carefully designed specimens free from notches or sharp changes in section, in a nominally non-corrosive atmosphere of ambient air, and the fatigue strength is taken as the maximum alternating load which a specimen will endure for a life of 10^7 cycles. In practice, the fatigue strength determined in this way can be related to the ultimate tensile strength (UTS) of a material and in many instances it is of the order of half the UTS. However, such tests are usually carried out on small specimens 1–2 cm in diameter and it has become evident that the actual size of the specimens has a bearing on the resultant fatigue strength, becoming lower the larger the specimen. This has to be taken into account when assessing the fatigue strength, and a factor of one third the UTS is more realistic.

Having chosen a material for a particular application the design has to be considered and the appropriate factors applied for any stress concentration introduced by the shape of the component. In practice it is found that many materials do not have their fatigue strength lowered to the expected value. This led to the concept of notch sensitivity.

1.1.2 Notch Sensitivity

Notch sensitivity (q) is defined as:

$$q = \frac{K_f - 1}{K_t - 1}$$

where K_t is the theoretical strength reduction factor and K_f is the strength reduction factor actually determined, i.e. the ratio of the fatigue strength of an unnotched specimen to that of a notched specimen. The maximum value that K_f can have is K_t and hence $q = 1$. If the notch has no effect then K_f is 1 and $q = 0$.

A material which is full of incipient notches already, such as grey cast iron where the graphite flakes have little or no strength, will not be affected by further notches and so q would be expected to be low or zero. Fully annealed materials with the capacity to work harden are also notch insensitive. On the other hand, high strength materials where the increased strength is the result of precipitation and ageing processes or work hardening have a much higher notch sensitivity. It also depends on the specimen size. Both these features are shown in Fig. 1.1.

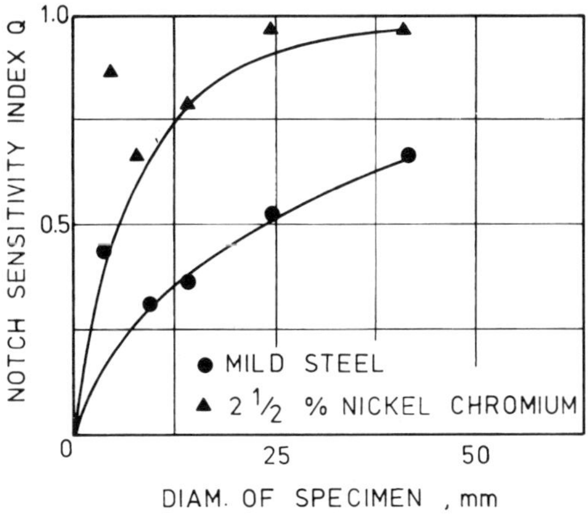

FIG. 1.1. Effect of material and specimen size on notch sensitivity.[1]

1.1.3 Environment

The second factor which can have a major effect on fatigue strength is the environment. The number of operating environments is countless if one considers the whole range of the chemical industry but in practice the most common corrosive environment is sea water. Most of the assessments of corrosion fatigue in common engineering materials have been made in sodium chloride solutions, usually 1% or 3%, or in a spray of such solutions. The effect on the *S*–log *N* curve is to cause it to fall much more steeply. Figure 1.2 illustrates this point. Since corrosion is a time-dependent chemical process, frequency is a much more significant variable. At low frequencies the corrosive attack per cycle will be greater unless the reaction is self-stifling. Figure 1.3 shows the effect of reducing the frequency on the fatigue behaviour of a carbon steel in 1% sodium chloride. Fortunately the effects of corrosion can usually be overcome by using corrosion resistant materials, although this inevitably involves increased expenditure, or by applying protective surface treatments such as plating, metal spraying, or diffusion coatings to a cheaper material which has the necessary mechanical properties.

1.1.4 Fracture Mechanics

More recently the trend has been to apply linear elastic fracture mechanics to the propagation of an existing crack. The stress intensity factor arising

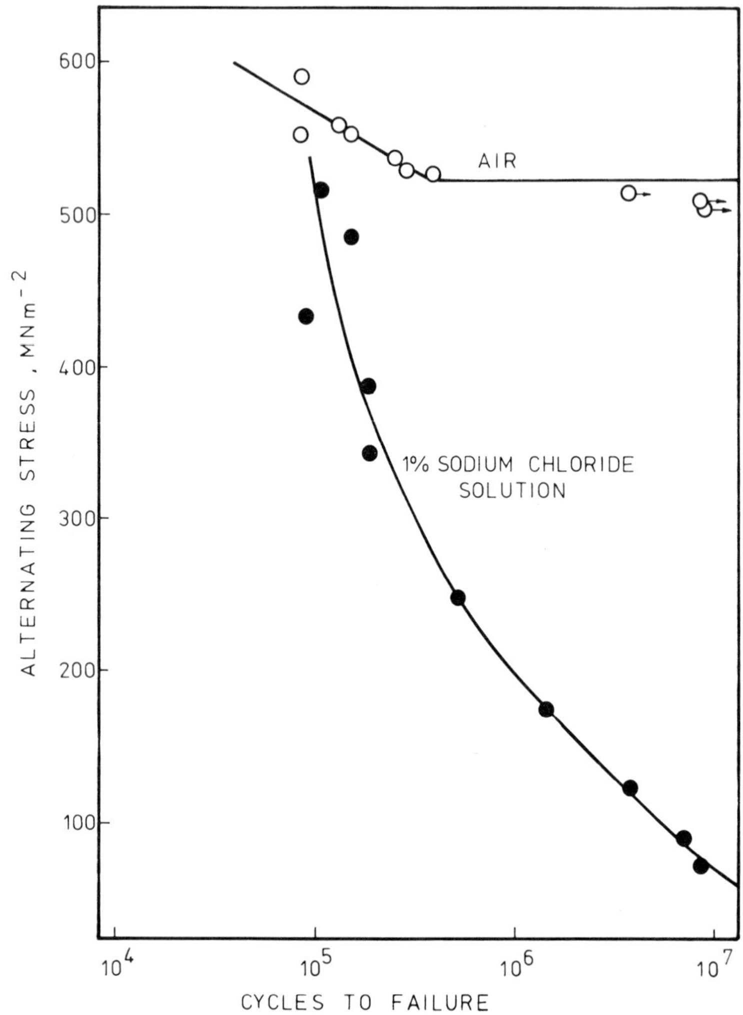

FIG. 1.2. Effect of environment on fatigue behaviour of 0·7C steel tested in rotating-bending.

from a crack of length a and an applied stress of σ is given by

$$K_{\mathrm{I}} = \sigma\sqrt{\pi a} \tag{1.1}$$

in the arrangement shown in Fig. 1.4.

The subscript I refers to the mode in which the surfaces of the crack are moving relative to one another. In this case the surfaces of the crack are

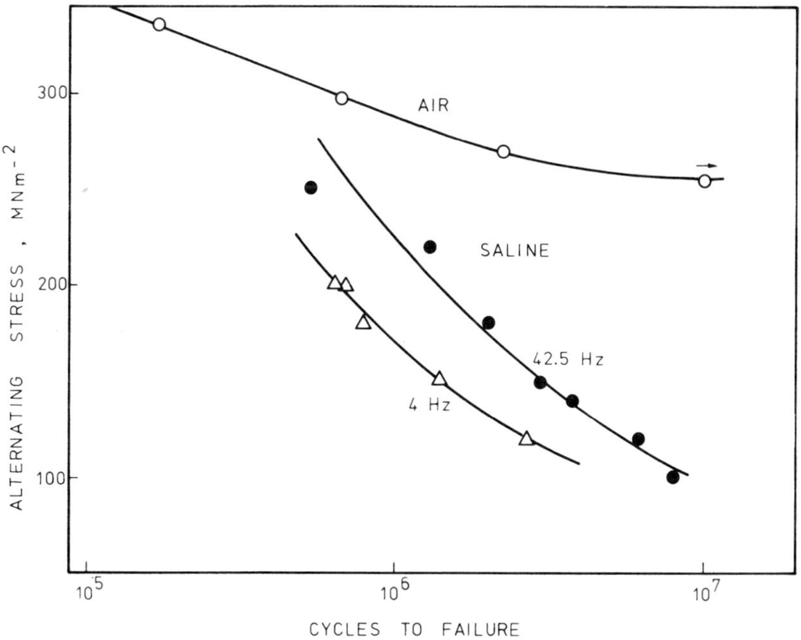

FIG. 1.3. Effect of frequency on fatigue behaviour of a carbon steel in saline solution.

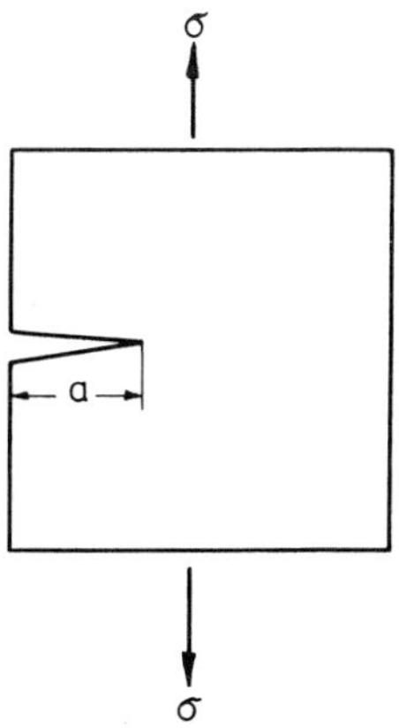

FIG. 1.4. Crack stressed in opening mode —mode I.

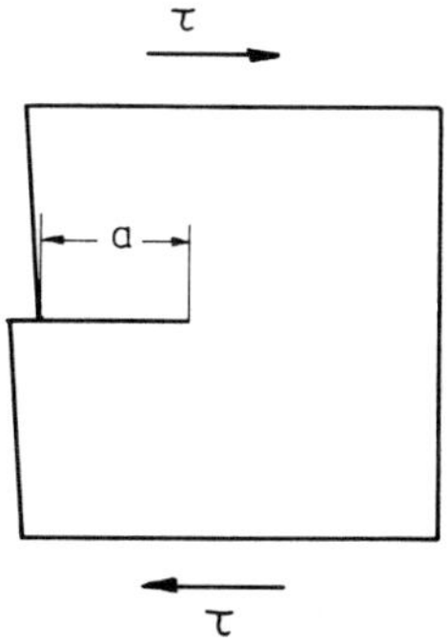

FIG. 1.5. Crack stressed in shear mode— mode II.

moving apart, i.e. the crack opening mode. If the applied stress were a shear stress, so that the surfaces of the crack were to slide over each other in a direction perpendicular to the leading edge of the crack, this would be mode II (Fig. 1.5).

This situation is the same as in the formation of an edge dislocation. The third mode is analogous to the formation of a screw dislocation. In practical studies of crack propagation mode I testing is used almost exclusively but mode II could be relevant to the case of fretting where an alternating shear stress is applied to the surface of the material. If a subsurface crack parallel to the surface is present as envisaged in the delamination theory the type of stressing would be mode II.

The actual stresses at any point in the vicinity of a crack tip are given by a set of equations of the form

$$\sigma \quad \text{or} \quad \tau = \frac{K}{\sqrt{2\pi r}} f(\theta) \tag{1.2}$$

where r and θ are the polar coordinates of the point. This is illustrated in Fig. 1.6 where the plane of the crack is the zx plane and the crack front is the x axis.

It is apparent that as r is decreased the stress will rise until at some point the yield stress of the material will be reached. When $\theta = 0$, $f(\theta) = 1$ so the point at which this occurs on the x axis is given by

$$r \simeq K^2/2\pi\sigma_0^2 \tag{1.3}$$

where σ_0 is the yield stress. Within this distance from the crack tip plastic deformation will occur which will modify the stresses in the surrounding elastic region, hence the relation is only approximate. Equation (1.3) gives the size of the plastic zone for plane stress, i.e. where $\sigma_z = 0$. This applies at the ends of the crack where it reaches the surface. In the middle of the specimen where σ_z is not zero a condition of plane strain obtains and the plastic zone is somewhat smaller, namely

$$r \simeq \frac{K^2}{6\pi\sigma_0^2} \tag{1.4}$$

In practice, thick specimens are used so that the testing is done under conditions of plane strain. As the crack grows, a critical value of K will be reached where the crack continues rapidly to complete failure. This critical value of K is termed the fracture toughness and designated K_{IC} when mode I, which is the most usual method used in testing, is applied. The units of K_{IC}

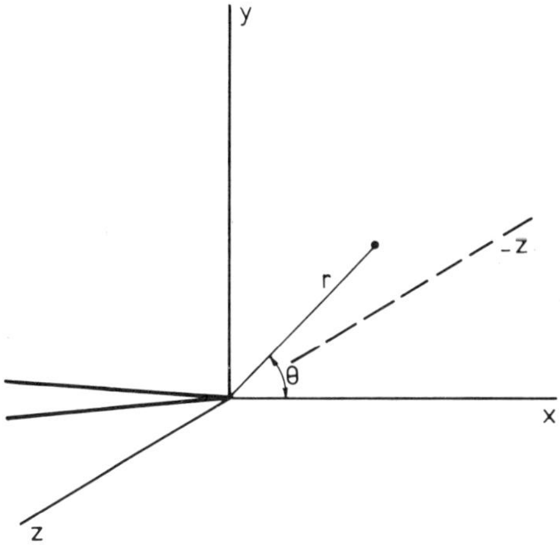

FIG. 1.6. Coordinate system at crack tip.[2a]

FIG. 1.7. Compact tension specimens.

are $MN\,m^{-3/2}$. Since the value of K_{IC} determined under plane strain conditions is independent of the thickness of the specimen, provided it is sufficiently thick for the conditions of plane strain to apply over a substantial distance of the crack front, it is a useful piece of design data for general application. Figure 1.7 shows two typical test pieces for such determination.

In making a determination of the plane strain fracture toughness using a specimen such as that in Fig. 1.7, a crack is generated from the notch by fatiguing the specimen. The progress of the crack is then monitored under the influence of a suitably applied stress either by determining the degree of opening of the crack, the compliance, or its length. The actual

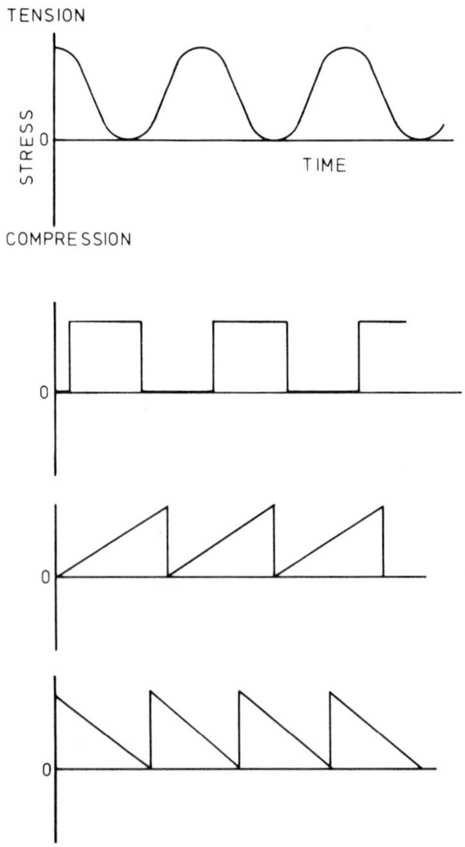

FIG. 1.8. Stress profiles used in crack propagation measurements.

determination of reliable values of K_{IC} requires the crack to be of a certain minimum length consistent with the dimensions of the specimen and the application of corrections for plastic deformation, but these details need not concern us here. The reader is referred to the detailed texts listed at the end of the chapter.

The application of fracture mechanics to fatigue is concerned with the growth of the crack under cyclic stressing. If the precracked specimen in Fig. 1.7 is cyclically stressed between two levels of tensile stress σ_{min} and σ_{max}, both tensile, then the stress intensity at the tip of the crack will be varying between K_{min} and K_{max}, both dependent on a the length of the crack. The difference between them, ΔK, will also be dependent on a. In practice σ_{min} is made zero or nearly zero so that the stress ratio $R\,(\sigma_{min}/\sigma_{max})$ is zero or nearly so. In the servohydraulic machines used for this type of testing the shape of the stress cycle can be chosen, i.e. sine wave, square wave, positive or negative sawtooth, etc. as shown in Fig. 1.8.

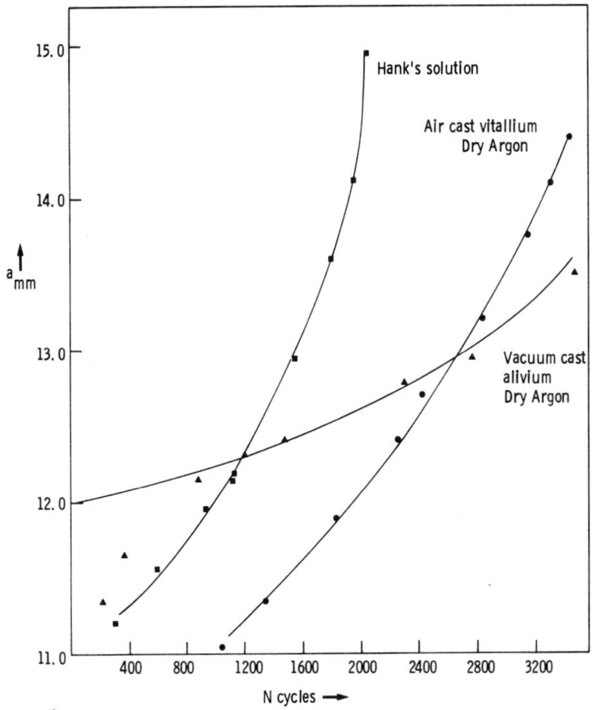

FIG. 1.9. Crack length v. number of cycles of stress for two cobalt-based alloys in different media.

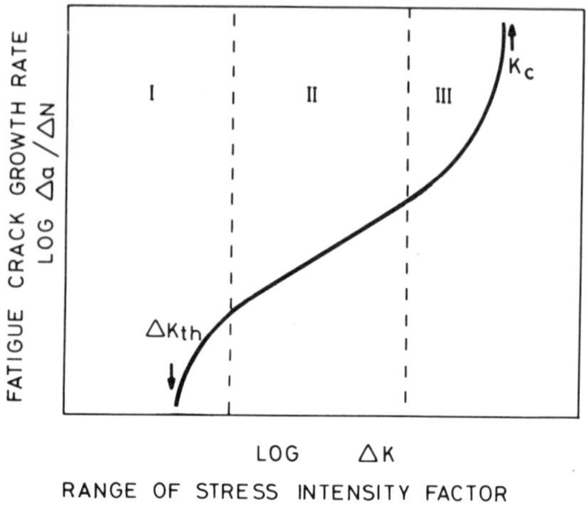

FIG. 1.10. Typical crack growth curve indicating a threshold value of stress intensity factor ΔK_{th}.

FIG. 1.11. Compact tension specimen surrounded by perspex cell to provide a flowing liquid or gas environment at the crack tip.

The growth of the crack is monitored either visually or by measuring the potential drop across the specimen and a curve is plotted of crack length against number of cycles as in Fig. 1.9. By taking tangents at various points on the curve and relating the slope to the stress intensity range ΔK at that crack length, a curve is plotted of log $\Delta a/\Delta N$ against log ΔK. The shape of the curve is shown in Fig. 1.10 and has three regions. In many systems there appears to be a threshold value of ΔK but determining crack rates at very low values of 10^{-8} or 10^{-9} mm per cycle at frequencies of 1 Hz or less is a very painstaking business. The second, linear, region is of greater interest and is the part of the curve to which the empirical Paris–Erdogan relation applies

$$\frac{\Delta a}{\Delta N} = C(\Delta K)^n$$

where C and n are constants. The exponent n usually has a value between 2 and 4. The third region is where the crack is accelerating to eventual failure.

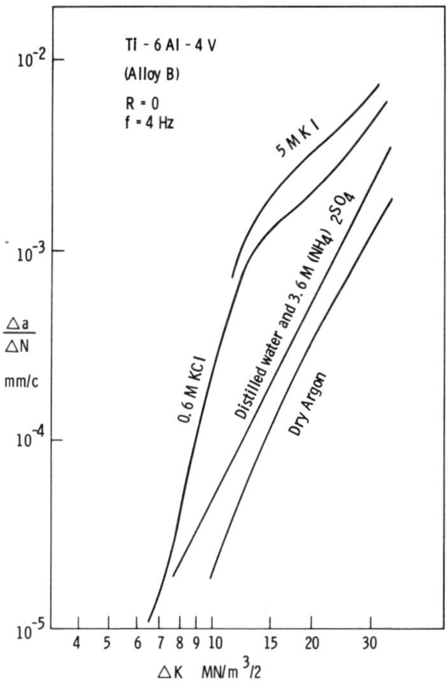

FIG. 1.12. Effect of various environments on crack growth rate in the titanium alloy Ti–6Al–4V at a frequency of 4 Hz and zero stress ratio.[2b]

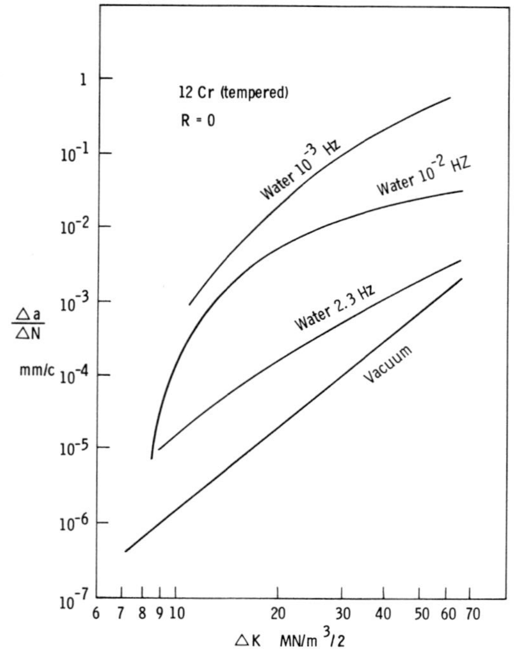

FIG. 1.13. Effect of frequency on crack growth rate in a 12 %Cr stainless steel tested in distilled water at zero stress ratio.[3]

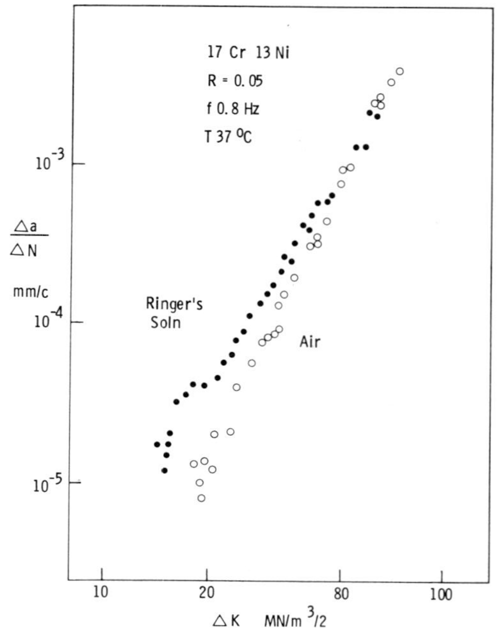

FIG. 1.14. Crack growth rate curves for an austenitic stainless steel in Ringer's solution (artificial body fluid) and air. Note enhanced environmental effect at low crack growth rates.[4]

Much recent work has been devoted to determining the effect of the environment on the linear portion of the curve. The progress of the crack in an inert medium, usually dry argon, is taken as the base line. This involves the design of a suitable cell to surround the specimen such as that shown in Fig. 1.11. Examples of the effect of various gaseous and liquid environments on the crack growth rate are shown in Fig. 1.12.

Since, as mentioned earlier, corrosion is a time-dependent process, frequency is an important variable as can be seen in Fig. 1.13. As is to be expected, the environment has a greater effect at the lower crack growth rates and has a big influence on the threshold value for crack growth (Fig. 1.14).

Some alloy systems are susceptible to stress corrosion cracking and this

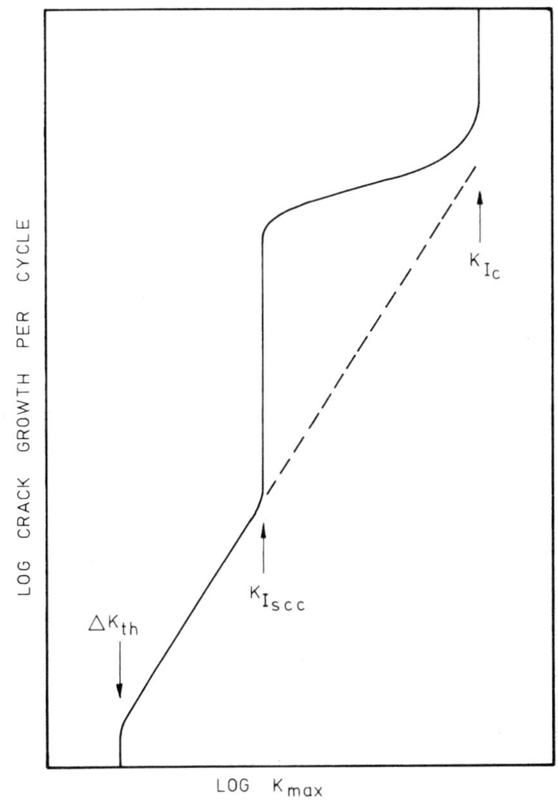

FIG. 1.15. Schematic representation of crack growth rate curve in a material susceptible to stress corrosion cracking.[2c]

phenomenon also has a threshold value of K associated with it, designated $K_{I_{scc}}$. If K_{max} exceeds $K_{I_{scc}}$ in the cyclic testing then a contribution to the crack growth is to be expected from this source and a sudden increase may occur (Fig. 1.15).

After such a test has been performed it is usual for the specimen to be broken apart and the portion of the fracture face relating to the propagation part of the curve to be examined optically or, more usually, in the scanning electron microscope. From this has developed the subject of fractography, i.e. the determination of whether the crack is propagated by cleavage or ductile tearing. It is at this stage that the physical influence of the environment can be recognised, but that aspect belongs to the next section.

1.2 MECHANISM OF FATIGUE

1.2.1 Initiation of the Fatigue Crack

Any discussion of fatigue must be related to the movement of dislocations under the influence of the applied stress. The threshold stress to cause a dislocation to move is very small, the so-called Peierls–Nabarro force, and therefore cyclic stressing well below the elastic limit will involve the oscillation of dislocations. It is only at the yield point that Frank–Read sources begin to operate and result in a massive increase in the number of dislocations and the consequent plastic deformation. The observation of persistent slip bands and extrusions at slip bands by Forsyth were explained in terms of the movement of dislocations on parallel or intersecting slip planes. Two examples are shown in Fig. 1.16. The development of thin foil electron microscopy allowed a more detailed description of dislocation patterns in fatigued crystals. The type of dislocation structures formed during fatigue depends on the stacking-fault energy of the material. A low stacking-fault energy makes cross-slip by dislocations from one slip system to another difficult and intersecting planar arrays of dislocations are formed, e.g. in the copper alloys. At high stacking-fault energies, cross-slip occurs leading to the formation of cellular structures, e.g. in aluminium. The relation between type of dislocation array and stacking-fault energy as a function of the number of cycles to failure is shown in Fig. 1.17.

One of the features associated with persistent slip bands which has been observed by some workers is the appearance of regularly spaced tubular holes. A similar observation has been made in a fretted area of a specimen of pure copper. In that case the tubular holes were nearly parallel to the surface due to the nature of the stressing (Fig. 1.18).

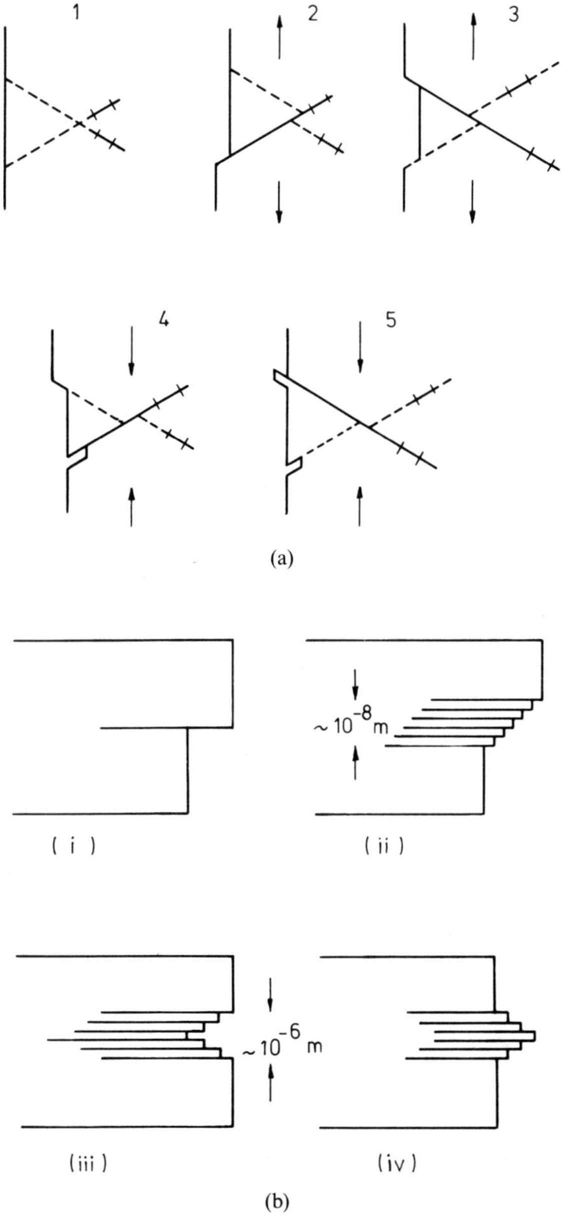

FIG. 1.16. Possible mechanisms for the formation of persistent slip bands on the surface of a test-piece.[5] (a) intersecting slip systems; (b) parallel slip systems.

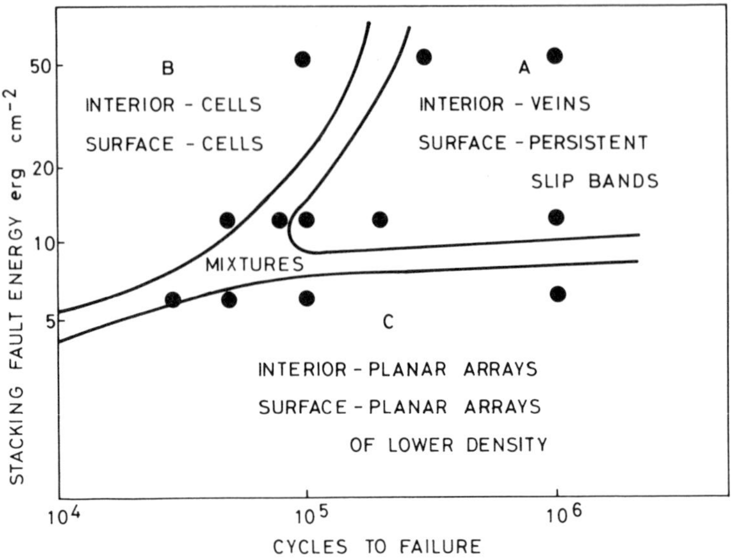

FIG. 1.17. Dislocation arrangements near the surface and within the material in a face-centred cubic metal as a function of stacking fault energy and stress amplitude—high stress amplitude = low number of cycles to failure.[2d]

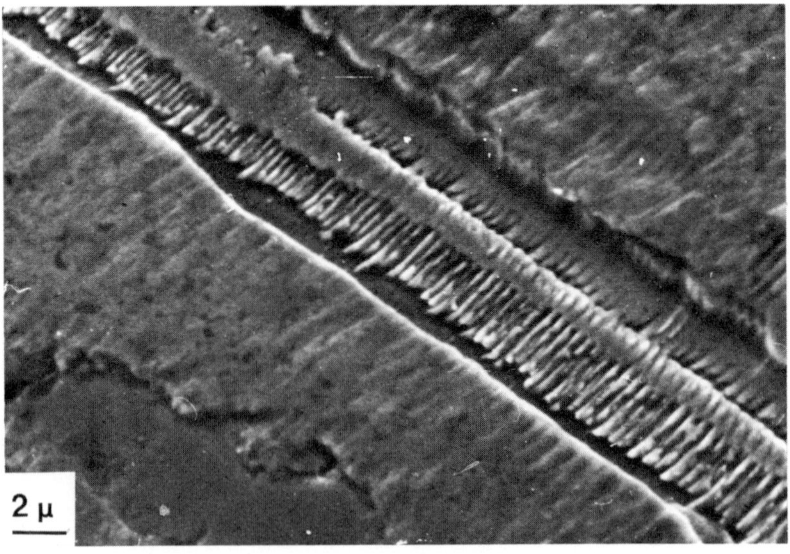

FIG. 1.18. Zip fastener effect produced by fretting on the surface of an oxygen-free high conductivity copper specimen.

1.2.2 Propagation of the Fatigue Crack

Initially the nucleated fatigue crack grows in the shear mode, i.e. at an angle of 45° to the applied stress, designated stage I. This is because slip occurs on the slip system where the resolved shear stress is a maximum, i.e. on the slip plane most nearly at 45° to the direct stress (Fig. 1.19).

The length of the stage I crack depends on the stress level and the environment. Low alternating stresses and a corrosive atmosphere

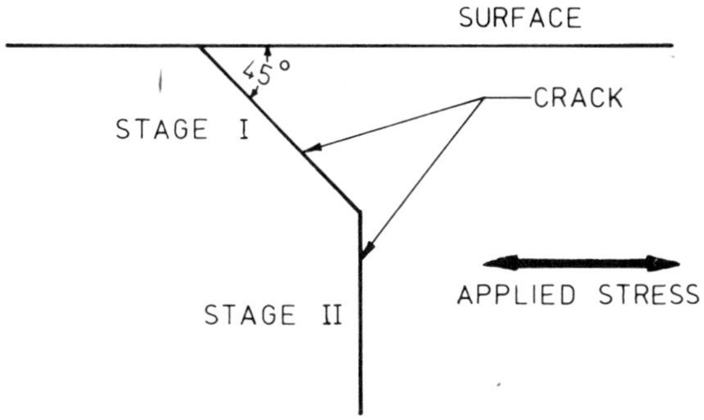

FIG. 1.19. Stage I and stage II development of a propagating fatigue crack.

encourage stage I growth. In stage II, the growth of the crack is dominated by the tensile stresses at the crack tip and the crack therefore proceeds in a direction perpendicular to the applied stress. The changeover from stage I to stage II is often at a grain boundary where the crack enters a crystal of different orientation to the one in which it was initiated.

In fretting fatigue the alternating shear stress applied to the surface by the fretting action usually encourages stage I growth to such an extent that a pronounced lip is formed on one fracture surface with a corresponding chip on the other (*see* Chapter 9, Fig. 9.9).

Examination of the fracture surface reveals that the stage II crack surface is relatively smooth in appearance but when examined under a high power microscope or the scanning electron microscope, striations are visible which correspond to the progress of the crack with each cycle of stress (Fig. 1.20). Each striation contains a portion of ductile failure and cleavage failure, the relative amounts depending on the overall ductility of the material and also the environment. With a reactive metal like aluminium the presence of water vapour can lead to the formation of Al_2O_3 and hydrogen on crack surfaces. Penetration by hydrogen encourages brittle or cleavage fracture. The formation of striations is illustrated in Fig. 1.21.

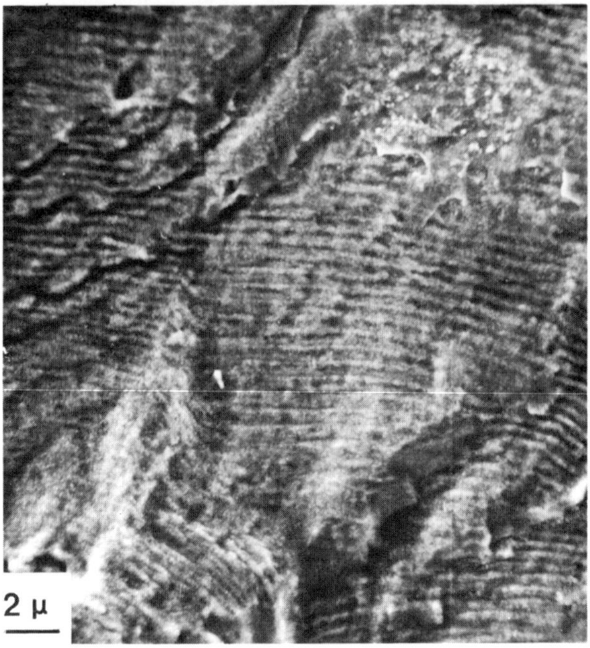

FIG. 1.20. Striations on fracture surface of a steel component which failed by fatigue in service.

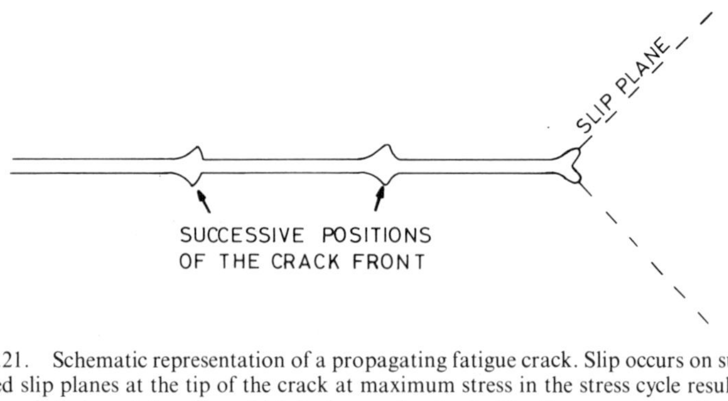

FIG. 1.21. Schematic representation of a propagating fatigue crack. Slip occurs on suitably oriented slip planes at the tip of the crack at maximum stress in the stress cycle resulting in creation of virgin metal surface which is vigorously attacked by the corrosive medium, with possible hydrogen penetration leading to embrittlement.[2e]

1.2.3 The *S*-log *N* Curve

The *S*–log *N* curve represents the end result of the initiation and propagation of a fatigue crack. However, some observations can be made concerning the influence on the fatigue curve when external factors such as environment or fretting have an effect on these two processes.

Any factor which lowers the fatigue limit from A to A′ in Fig. 1.22 is having an influence on the initiation of the crack since at applied stresses between A and A′ a crack is formed which propagates to failure, whereas in the absence of the factor no failure would occur. On the other hand, any

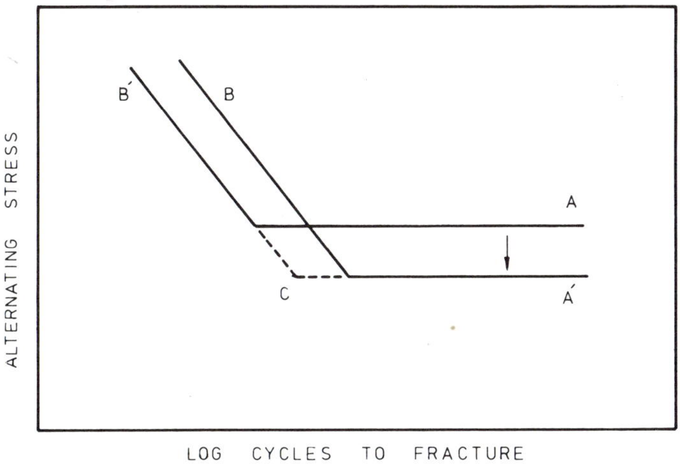

FIG. 1.22. Schematic representation of the *S*–log *N* curve.

factor which shifts the upper part of the curve to the left, i.e. from B to B′ is having an effect on the propagation of the crack since the life is shorter at any given stress. In practice, both parts of the curve are often affected and the result is the curve B′CA′. Figure 1.23 shows the effect of a phosphate additive in a white mineral oil on the fatigue curve of a 3% nickel steel. In this case the fatigue limit remains unchanged but the upper part of the curve is shifted to the left, implying that the additive has an effect on the propagation but not on the initiation of the fatigue crack.

Fretting is particularly effective in initiating a fatigue crack but once it has grown out of the immediate region of the surface where the additional shear stress is acting, the propagation rate will be little influenced. This is illustrated in Fig. 1.24 which shows the effect of fretting on the fatigue behaviour of an 18Cr 10Ni austenitic steel.

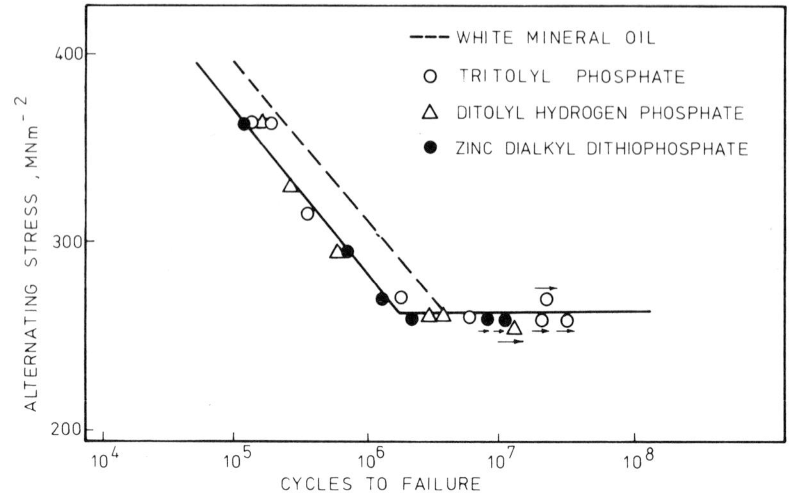

FIG. 1.23. Effect of various additives to a white mineral oil on the fatigue behaviour of 3 % Ni steel.[6]

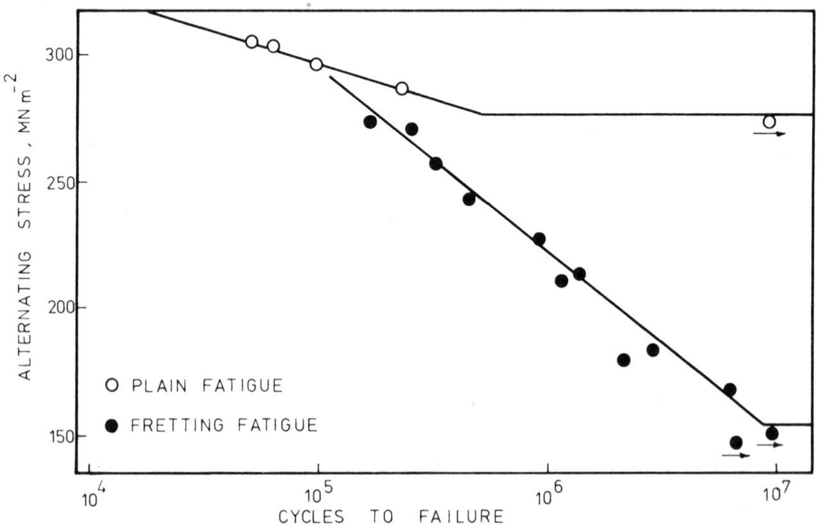

FIG. 1.24. Effect of fretting on the fatigue behaviour of an austenitic steel. At stresses just above the fatigue limit, the fatigue life is similar for plain fatigue and fretting fatigue, indicating that when a crack is initiated, whether by fretting or not, its propagation rate is approximately the same in both cases.

1.3 CONCLUSION

This brief review is meant to introduce a number of topics which are relevant to the occurrence of fretting fatigue. Those of particular interest are (a) notch sensitivity since notch-sensitive materials suffer large reductions in fatigue strength due to fretting, (b) fracture mechanics since its application can lead to predictions of fretting fatigue behaviour, and (c) environmental effects as these have a profound influence on the severity of the damage.

These specific questions are dealt with in much greater detail in the chapters which follow, but a list of relevant texts is given below for the reader who wishes to study the subject of fatigue in more depth.

REFERENCES

1. Pope, J. A. *Metal fatigue*, Chapman and Hall, London, p. 56 (1959).
2. Devereux, O. F., McEvily, A. J. and Staehle, R. W. *Corrosion fatigue*, Nat. Assoc. Corr. Engrs., Houston (1972) *a* p. 15; *b* p. 333; *c* p. 412; *d* p. 96; *e* pp. 305 and 388.
3. Speidel, M. O. *Stress corrosion cracking and hydrogen embrittlement of iron alloys*, Nat. Assoc. Corr. Engrs., Houston, p. 1082 (1977).
4. Wheeler, K. R. and James, L. A. *J. Biomed. Mater. Res.*, **5**, 277 (1971).
5. Frost, N. E., Marsh, K. J. and Pook, L. P. *Metal fatigue*, Clarendon Press, Oxford, p. 29 (1974).
6. Galvin, G. D. and Naylor, H. *Proc. Inst. Mech. Engrs.* **179**(3J) 60 (1964–65).

BIBLIOGRAPHY

1. Broek, D. *Elementary engineering fracture mechanics*, Noordhoff International Publishing, Leyden (1974).
2. Frost, N. E., Marsh, K. J. and Pook, L. P. *Metal fatigue*, Clarendon Press, Oxford (1974).
3. Pope, J. A. *Metal fatigue*, Chapman and Hall, London (1959).
4. McEvily, A. J. and Staehle, R. W. *Corrosion fatigue: chemistry, mechanics and microstructure*, Nat. Assoc. Corr. Engrs., Houston (1972).

2

The Role of Elastic Stress Analysis in the Interpretation of Fretting Fatigue Failures

J. J. O'CONNOR
St. Peter's College, Oxford University, UK

2.1 INTRODUCTION

Fretting damage occurs on the contacting surfaces of components which are clamped together and which, although nominally at rest relative to each other, are subject to vibration. It is caused by repeated alternating slipping of the surfaces over part of their mutual contact area. Visible damage is found when the amplitude of slip is as little as 10^{-4} mm.

In this chapter, the contribution which elastic stress analysis can make to the understanding of fretting and fretting fatigue is discussed. The chapter starts with a review of those configurations for which algebraic solutions have been found, i.e. the contact of spheres and cylinders under a steady normal load and an alternating frictional shear load. For more complicated geometries, a method of analysis based on finite elements is described. A simple model of a lap joint loaded in alternating tension or bending and axially-symmetric shrink-fitted assemblies loaded in alternating torsion is considered. The results of fatigue tests on such assemblies are discussed in the light of the calculated stress fields and the patterns of slip and separation of the contact surfaces.

2.2 CONTACT OF SPHERES AND CYLINDERS

Firm evidence of a direct correlation between the incidence of slip and the onset of fretting followed from studies of the contact of elastic spheres. Cattaneo[1] and Mindlin[2] extended the Hertz solution in the theory of elasticity for the contact of two spheres under a normal compressive force to account for the addition of a shearing force less than limiting friction applied parallel to the contact surface, Fig. 2.1a. Mindlin started by

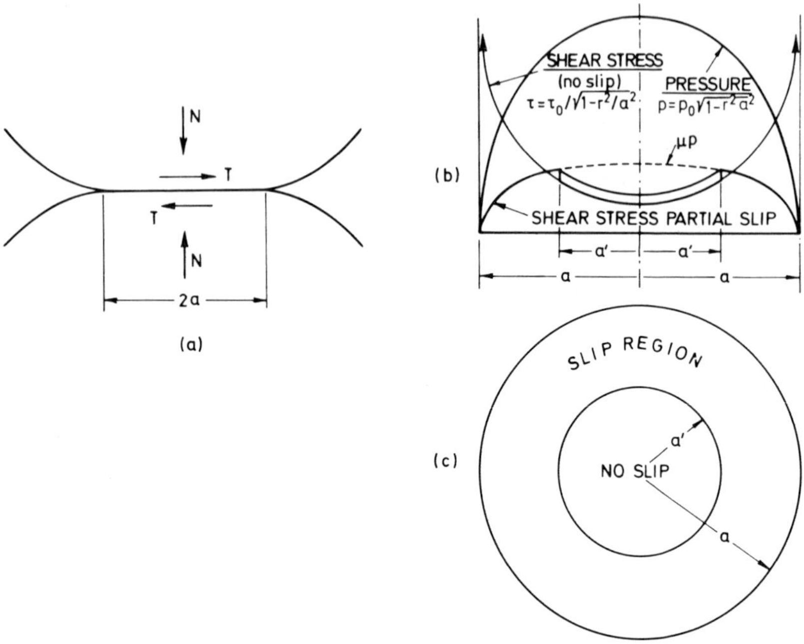

FIG. 2.1. (a) Contact of elastic spheres under normal and tangential forces. (b) Distribution of pressure on the contact surfaces according to Hertz. (c) Plan view of contact surface. Suppression of interfacial slip requires infinite shear traction at the edge of the contact region. A finite frictional capacity μp implies slip on the annulus $a' < r < a$ with the central region $r < a'$ remaining unslipped.

assuming that the interface had an infinite frictional capacity so that slip could not occur and demonstrated that shear tractions rising to infinity at the edge of the contact region would then be required, Fig. 2.1b. The solution was modified by assuming that the shear traction at any point on the interface could never exceed limiting friction there. Limiting friction at any point was taken to be given by the product μp of a constant coefficient of friction μ and the local value p of the Hertzian pressure. It was then found that the interface is divided into two distinct regions, a circular region $r \leq a'$ where the surfaces remain locked together and an annular region $a' < r \leq a$, extending to the edge of the contact area, over which slip occurs, Fig. 2.1c. The dimensions of the central locked region were given as

$$\frac{a'}{a} = \left[1 - \left(\frac{T}{\mu N}\right)\right]^{1/3} \tag{2.1}$$

where N is the value of the clamping force and T the shearing force. When limiting friction is reached overall, $T/\mu N \to 1$, the radius of the locked region shrinks to zero and macroscopic sliding commences. If the shearing force is caused to alternate between the limits $\pm T^*$, $T^* < \mu N$, the region $r < a'$ remains locked but the direction of slip in the annular region also alternates.

Johnson[3] provided convincing experimental evidence in support of various aspects of this theory. The contact surfaces of metal spheres which had been clamped together by a steady load and subjected to alternating shearing forces were also divided into two regions, a central circular region where the surfaces remained undamaged and an outer annular region heavily fretted. The dimensions of these regions accorded well with Mindlin's result, small discrepancies being consistent with uncertainties about the value of μ and the possibility that repeated alternating slip could give rise to a steady growth of the coefficient of friction in the annular fretted region. O'Connor and Johnson,[4] and Goodman and Brown[5] found further evidence that the coefficient of friction grows with repeated alternating slip.

Johnson[3] calculated the amplitude of slip throughout the annular region. The maximum value α occurs at the edge of the contact area and has the value

$$\frac{\alpha}{a} = \frac{\pi(2-\nu)}{8} \cdot \frac{\mu p_0}{G} \left\{ \left[1 - \frac{2}{\pi} \sin^{-1}\left(\frac{a'}{a}\right) \right] \left[1 - 2\left(\frac{a'}{a}\right)^2 \right] + \frac{2}{\pi}\left(\frac{a'}{a}\right)\left[1 - \left(\frac{a'}{a}\right)^2\right]^{1/2} \right\} \quad (2.2)$$

where $p_0 = 3N/2\pi a^2$ is the maximum Hertzian pressure, and G and ν are the shear modulus and Poisson's ratio of the material respectively. The amplitude of slip at the edge of the contact area increases with increasing T until, at the point of incipient sliding when $T/\mu N \to 1$ and $a'/a \to 0$,

$$\frac{\alpha}{a} = \frac{\pi(2-\nu)}{8} \cdot \frac{\mu p_0}{G} \quad (2.3)$$

Even at the point of incipient sliding, the maximum slip amplitude is several orders of magnitude smaller than the radius of the contact circle. It can be seen why some authors use the term 'micro-slip'.

These results demonstrate the role which elastic stress analysis can play in the discussion of the fretting problem. As well as the elastic stress fields within the contacting solids, the extent of the slip region and the amplitude of slip can be determined. Indeed, Mindlin's approach suggests that contact surfaces slip in order to relieve otherwise high concentrations of shear stress on the interface, an effect entirely analogous to the onset of plastic deformation in ductile materials. From this point of view, slip and fretting in clamped joints may be thought to have a beneficial effect! A further beneficial effect arises from the energy dissipated against friction. Such energy dissipation gives to the joint a damping capacity useful in controlling vibration and is an important source of structural damping.

There are, however, concomitant deleterious effects. The contact surfaces are damaged by fretting and fatigue cracks are often found to propagate from the damaged portion of the interface, the phenomenon of fretting fatigue. Some insight into the mechanisms of crack initiation and propagation may be gained by examining the stress field in the Hertz/Mindlin problem and, in particular, by looking for tensile stress.

Hamilton and Goodman[6] calculated the state of stress for the limiting case of incipient sliding, $T/\mu N \rightarrow 1$. Some of their results are given in Fig. 2.2 in which is plotted the normal stress σ_x in the plane of the surface along a line through the centre of the contact circle parallel to the shearing force T. The curve in Fig. 2.2a shows the effect of the compressive force acting alone. The surface is in a state of compression within the contact area and tension without. The peak tensile stress occurs at the edge of the contact area and has a value of $[(1 - 2\nu)/3]p_0$, about $0{\cdot}13\,p_0$. The tensile stresses in the surface arise because the compressed material within the contact region tends to drag the material outside radially inwards. These tensile stresses are thought to be responsible for the ring cracks formed during Hertzian indentation of glass.

The application of a shear force tends to put the material ahead of the contact into compression and the material behind into tension. Figure 2.2b shows the distribution of normal stress σ_x along the same line when sliding is imminent at $\mu = 0{\cdot}25$ and $0{\cdot}5$. The maximum compressive stress is still found within the contact circle but at a point some way ahead of its centre. Hamilton and Goodman showed that the stress at the leading point of the contact circle, $x = +a$ becomes compressive for $\mu > 0{\cdot}079$. The stress σ_x rises to a sharp tensile peak at the trailing point $x = -a$, having a value there of $p_0\{[(1 - 2\nu)/3] + [\pi(4 + \nu)/8]\mu\}$, increasing directly with μ. Thus, if the shearing force alternates through the range $\pm\mu N$, the stress σ_x at $x = \pm a$ oscillates through a range $\pm\mu p_0 \cdot [\pi(4 + \nu)/8]$ about a mean tensile value of $(1 - 2\nu)p_0/3$.

Furthermore, Hamilton and Goodman located the point at which yielding would first occur in accordance with the von Mises criterion. For μ

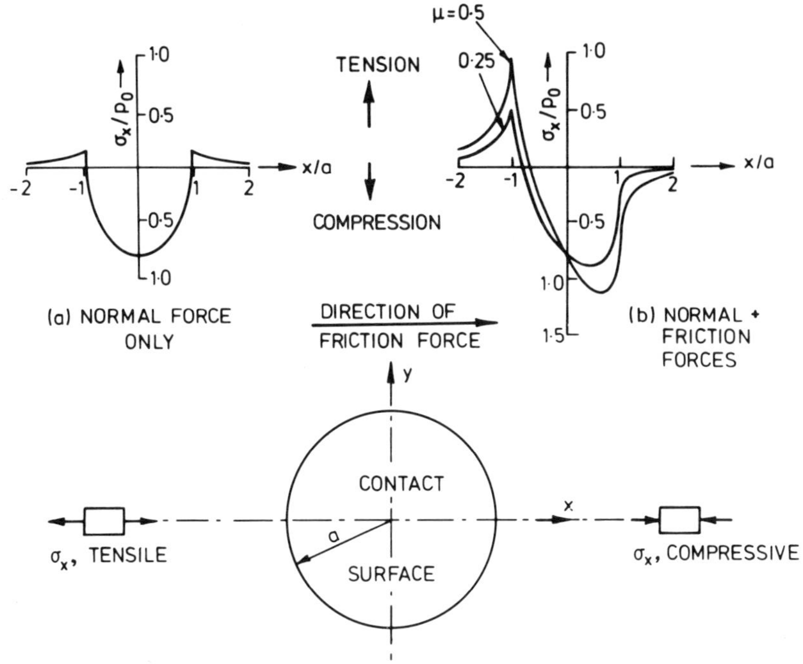

FIG. 2.2. Contact of elastic spheres. Distribution of normal stress σ_x in the plane of the contact surface along $y = 0$. (a) under normal load only and (b) at the instant of imminent sliding.

less than about 0·3, yield begins below the contact surface but for larger values of the coefficient of friction, yield first occurs on the surface, again at $x = -a$. With each reversal of the shearing force, localised yielding at the surface could lead to the initiation of cracks by a process of extrusion and intrusion of slip bands such as that described by Cottrell or Mott. The oscillating stress σ_x parallel to the surface would then be available to propagate such a crack into the bulk of the material.

Very similar conclusions follow from a two dimensional study of the contact of elastic cylinders under normal and shearing forces.[7] In this case, the stress σ_x at the edge of the contact strip alternates through a range $\pm 2\mu p_0$ and initial yield again occurs at the surface of the edge of the contact strip for sufficiently large μ. A view of the stress field under a cylindrical contact is shown in Fig. 2.3. Contact occurs over $-a < x < a$. Under the normal load alone, both principal stresses are compressive everywhere. The contour $\sigma_2 = -0{\cdot}1\,p_0$, normal load only, indicates how the load diffuses

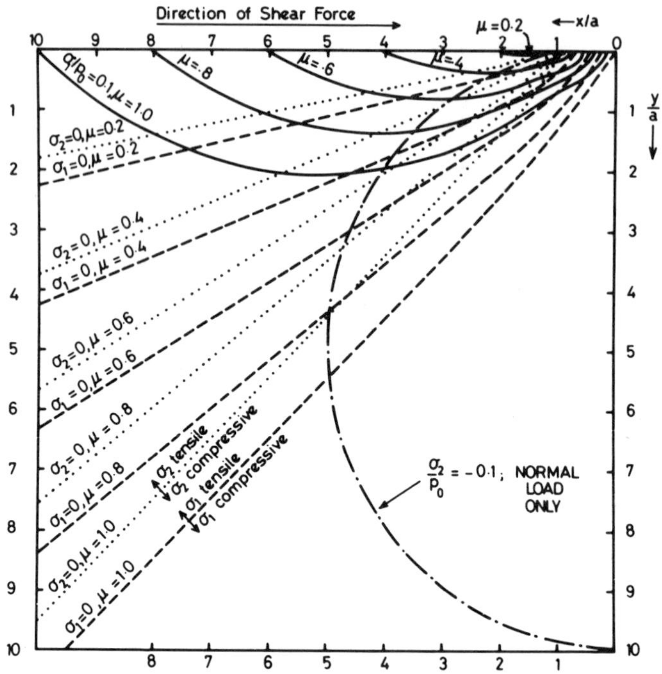

FIG. 2.3. Stress field under a sliding cylindrical indentor. Under normal load only, the largest compressive principal stress exceeds $0{\cdot}1\,p_0$ in magnitude in a circular region of area $49\pi a^2/2$. At the point of sliding, the principal stresses are tensile above the lines $\sigma_1 = 0$, or $\sigma_2 = 0$. The largest tensile principal stress exceeds $0{\cdot}1\,p_0$ in magnitude above the solid contours.

into the bulk from the surface and gives a measure of the volume of material influenced by the contact. When the shearing force is added, both principal stresses become tensile behind the contact in an approximately triangular region stretching down from the surface. The area of the triangular region grows directly with increasing μ. The contours along which the larger tensile principal stress has the value $0{\cdot}1\,p_0$ are shown for various values of μ, demonstrating that an increasing volume of material is subjected to appreciable tension as the coefficient of friction grows.

The areas enclosed by the contours of constant σ_1/p_0 as a proportion of $\pi a^2/2$ are shown in Fig. 2.4 plotted against μ. The graphs give an indication of the depth below the surface in which conditions would be favourable for the propagation of a crack initiated at the surface. If a crack were propagated by the contact stress field alone, its progress would be liable to be impeded as it approaches the zone of purely compressive stress.

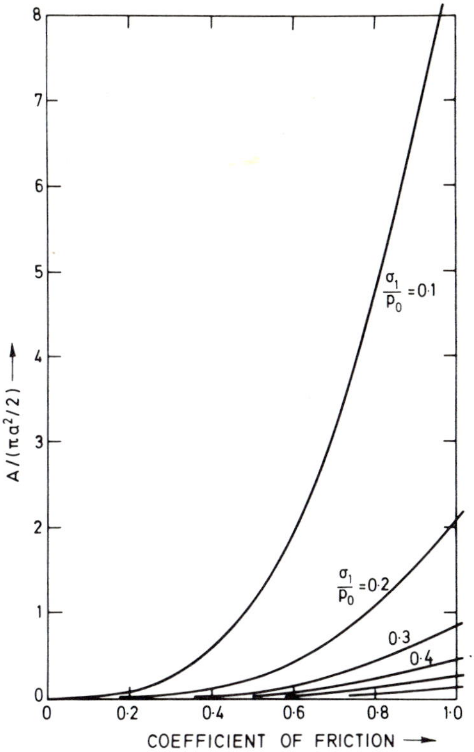

FIG. 2.4. Magnitude of the area under a sliding cylindrical indentor in which the maximum tensile stress exceeds $0{\cdot}1p_0$, $0{\cdot}2p_0$ etc. as a function of the coefficient of friction.

However, as the coefficient of friction grows, the amount of material affected by tensile stresses in excess of a given value grows very rapidly. For example, as μ grows from 0·2 to 1·0, the area bounded by the $\sigma_1 = 0{\cdot}1\,p_0$ contour grows by a factor of 220. The depth of crack which can propagate from a frictional contact might be expected to increase with μ in a similar manner.

In discussing a mechanism for the initiation and propagation of cracks from a frictional contact, it has not been necessary to make a direct appeal to the possible contribution to these processes of the surface damage caused by fretting. On the one hand, fretting results from repeated slip which occurs in order to relieve otherwise large concentrations of shear stress on the contact surface. On the other hand, the stress field set up by an alternating frictional contact creates conditions favourable for the initiation and propagation of fatigue cracks. The question then arises whether fretting and fatigue are cause and effect or could both phenomena be thought to be effects of the same common cause, the contact stress field?

A similar question arises when seeking an explanation for the fatigue failure of assembled machine and structural components. The assemblies in Fig. 2.5 may be typical. The clamping load is represented by the force N.

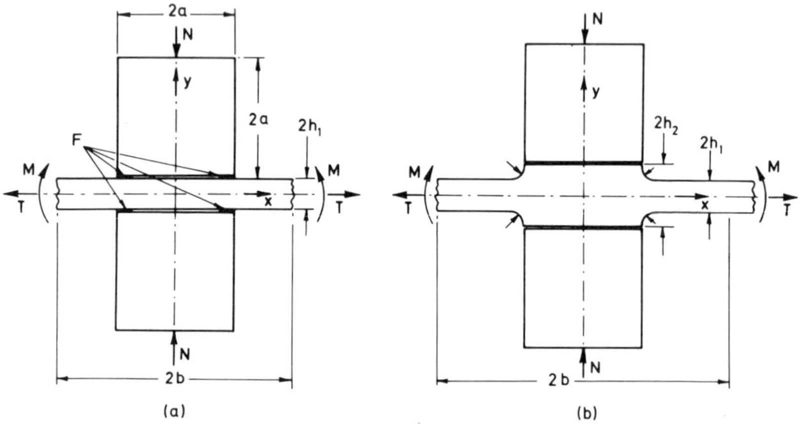

FIG. 2.5. (a) The 'sharp' assembly. (b) The 'round' assembly. In all the calculations reported here, $a/h_1 = 4$, $b/h_1 = 8$, $h_2/h_1 = 2$.

The assemblies can be loaded in alternating tension (push–pull) or bending or, if axially symmetric, torsion. Two features of the behaviour of the 'sharp' assembly, Fig. 2.5a, are of interest. Fatigue cracks, leading to failure of the central bar, propagate from near the mouth of the assembly. Fretting damage is found on the contact surfaces in the regions marked F on the diagram. In discussing the failure of components of this shape, some authors ascribe the initiation and propagation of cracks to the stress concentration which would be expected at the sudden change in section; others, when faced with the fretted contact surfaces, ascribe the failure to fretting fatigue. While considerable improvements in strength may be obtained by using a design similar to the 'round' assembly of Fig. 2.5b, the reason for the improvement may not be altogether clear. On the one hand, the geometric stress concentration should be much reduced by the introduction of the transition section; on the other, so might be the intensity of fretting on the interfaces.

Indeed, if the interfaces in such assemblies slip in order to relieve even larger stress concentrations, it might be argued that the fatigue strength of a given assembly would be higher than that of a geometrically similar solid made of one piece of material.

An assessment of the cause of failure of a component such as those of Fig. 2.5 requires a knowledge of the stress-field within the assembly. The stress analysis should allow for the possibility of slip against finite friction, and could yield estimates of the extent of the slip region and the amplitude of slip. Further, the stress analysis should allow for the possibility of partial separation of the contact surfaces under load, shown in an exaggerated form in Fig. 2.6. In the next section, a numerical method is described, based

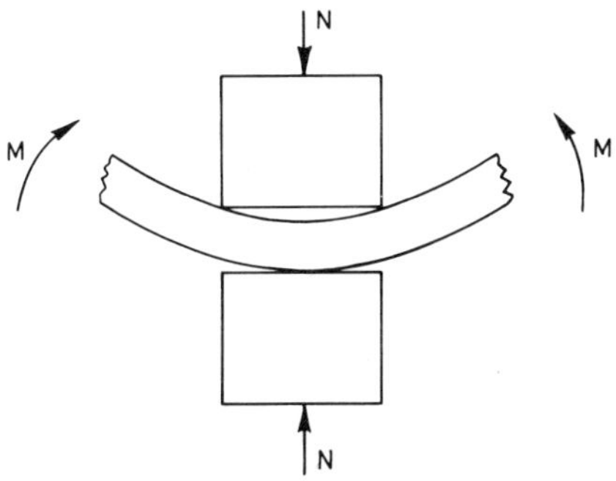

FIG. 2.6. Exaggerated view of separation of the contact surfaces under bending loads.

on the finite element method, which makes allowance for slip and separation. This type of analysis yields the macroscopic distribution of stress in the assembly but, by assuming the surfaces to be geometrically smooth and the contact to be continuous, it makes no allowance for surface roughness generated by machining processes. However, it would not be unreasonable to assume that the microscopic stress fields associated with the contact of the asperities on metallic surfaces are described approximately by the contact of elastic spheres and cylinders discussed in this section. A superposition of the macroscopic and microscopic solutions would then be required.

2.3 STRESS ANALYSIS METHOD

Wright[8] and Wright and O'Connor[9,10] developed the method of analysis and applied it to the configurations of Fig. 2.5 loaded by a steady clamping force N and an alternating tensile force $\pm T^*$. Bramhall[11] extended the method to the same configurations when loaded by alternating pure bending moments $\pm M^*$. In this section, we shall describe the formulation of the bending problem, the simplifications to be introduced to recover the formulation for push–pull loading and indicate the further extensions made by Lau[12] needed for the study of axially-symmetric versions of those configurations loaded in alternating torsion.

2.3.1 Statement of the Problem

The system under consideration consists of a plate, against the plane surfaces of which are pressed two identical rectangular blocks under equal co-linear forces N under conditions of plane strain. A pure bending moment M is applied on sections outside the interface, as indicated in Fig. 2.5a. Assuming that the system is symmetric about the plane $x = 0$, only half the configuration need be considered, with the plane $x = 0$ remaining plane. The object is to determine the distribution of stress, the location of the slip and separation regions on the interfaces, the amplitude of the slip and separation displacements when, with the forces N held constant, the bending moment is allowed to alternate about zero with amplitude M^*.

It will be assumed that the interfaces have a limiting strength in friction, with a constant coefficient of friction, and have no strength in adhesion. Although the contacting bodies are considered to be linearly elastic, these assumptions render the problem non-linear in the sense that the current state of stress depends on the history of loading.

There are three types of region on each interface requiring three sets of boundary conditions. The index $i = 1, 2$ is used for the upper and lower interfaces, respectively. In the non-slip regions, the surface displacements u_i, v_i and U_i, V_i of the blocks and the plate respectively must be continuous and the shear stress τ at any point must be less than limiting friction μp, where μ is the coefficient of limiting friction and p is the normal pressure at that point. In the slip regions, the normal surface displacements must be continuous whilst the shear stress must equal limiting friction, the direction of the shear stress opposing the direction of slip. In the separation regions, both components of the surface traction must be zero. Displacements u_i, v_i are measured relative to origins 0_i in the blocks whilst the displacements U_i, V_i are measured relative to the origin $(0, 0)$ in the plate, which is considered to be fixed (Fig. 2.7).

The origins 0_i approach $(0, 0)$ by distances Δ_i under the combined action of the loads N and M, the distances Δ_i differing from each other because the system is not symmetrical about the plane $y = 0$. Using the sign

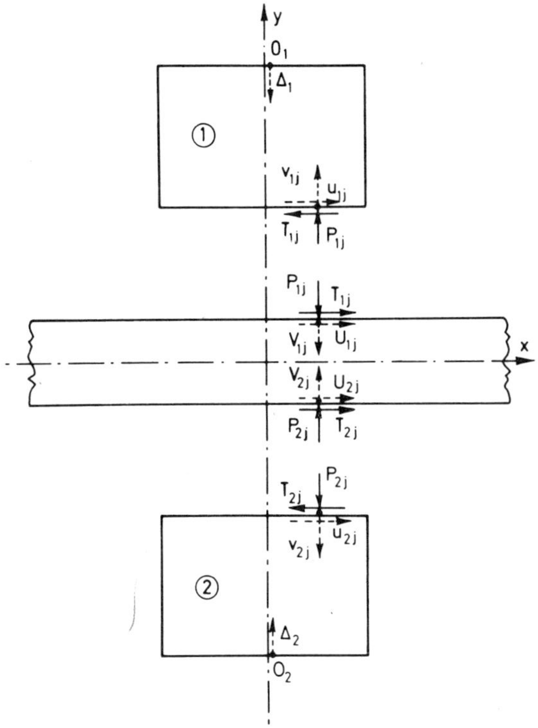

FIG. 2.7. Sign convention and notation for positive forces and displacements. The subscript 1 refers to quantities in the upper half, $x > 0$, of the system, and subscript 2 refers to the lower half, $x < 0$.

conventions for positive displacements and stresses shown in Fig. 2.7, the boundary conditions may be written as follows:

(a) For the non-slip regions

$$v_i(x) + V_i(x) = \Delta_i \qquad i = 1, 2 \tag{2·4}$$

$$u_i(x) - U_i(x) = 0 \qquad i = 1, 2 \tag{2.5}$$

with the subsidiary conditions

$$|\tau(x)| < \mu p(x) \qquad p(x) > 0 \tag{2.6}$$

(b) For the slip regions

$$v_i(x) + V_i(x) = \Delta_i \qquad i = 1, 2 \tag{2.7}$$

$$|\tau(x)| = \mu p(x) \tag{2.8}$$

with the subsidiary conditions

$$\mathrm{Sgn}[u_i(x) - U_i(x)] = \mathrm{Sgn}[\tau(x)] \qquad p(x) > 0 \tag{2.9}$$

(c) For the separation regions

$$p(x) = 0 \tag{2.10}$$
$$\tau(x) = 0 \tag{2.11}$$

with the subsidiary condition

$$v_i(x) + V_i(x) > \Delta_i \qquad i = 1, 2 \tag{2.12}$$

Equilibrium of the blocks in the y-direction is ensured if

$$\int_{-a}^{a} p_i(x)\,\mathrm{d}x = N \qquad i = 1, 2 \tag{2.13}$$

where $2a$ is length of the contact surface.

In order to solve these equations for the tractions τ and p, the influence functions relating surface displacements and tractions for each component of the system must be calculated in order to express all the boundary conditions in terms of τ and p. The bending moment M is introduced through the relations between U_i, V_i and M, so that M then appears explicitly in eqns (2.4), (2.5) and (2.7). At the outset, for specified N and M, the displacements Δ_i are both unknown and form part of the solution. In addition, the locations of the no-slip, slip and separation regions are unknown, as are the slip amplitude $\alpha_i(x) = u_i(x) - U_i(x)$ and the separation amplitude $\beta_i(x) = v_i(x) + V_i(x) - \Delta_i$. However, the tractions must be continuous at the boundaries between the various regions whilst, from symmetry, $\tau(x), x = 0$, must be zero.

2.4 METHOD OF SOLUTION

An entirely numerical method was used. The traction distributions were replaced by $2n$ concentrated normal compressive forces P_{ij} $(i = 1, 2; j = 1, 2 \ldots n)$, and $2(n-1)$ concentrated tangential forces T_{ij} $(i = 1, 2; j = 2, 3 \ldots n)$, at n points on each interface (Fig. 2.7). Influence matrices for each of the components of the system were obtained by standard finite-element methods,[13] replacing the elastic continua with meshes of triangular elements, each connected at six nodes to its neighbours, the nodes on the contacting surfaces being co-incident in the unloaded state. The elements of the influence matrices were calculated by obtaining the surface displacement response at each interface node on each component due to unit positive concentrated normal and tangential forces applied successively at each of the interface nodes on that component. In addition, the surface displacements U_{ij}, V_{ij} $(i = 1, 2, j = 1, 2 \ldots n)$ of the interface nodes on the plate due to a unit bending moment applied by a linear stress distribution $\sigma_x = -(3y/2h_1^3)$ on the section $x = \pm b$ (Fig. 2.5) were calculated (where h_1 is the semi-thickness of the plate outside the interface).

The boundary conditions, eqns (2.4), (2.5), (2.7), (2.8), (2.10), (2.11) and (2.13), now consist of sets of linear simultaneous equations on the unknown forces P_{ij}, T_{ij}. However, since the locations of the slip and separation regions were not known in advance, an iterative method of solution was used. Assuming initially that there was no slip and no separation, eqns (2.4) and (2.5) were solved and the solution inspected for values of $P_{ij} < 0$ and values of $|T_{ij}| > \mu P_{ij}$. For those nodes for which $P_{ij} < 0$, eqns (2.4) and (2.5) were replaced with eqns (2.10) and (2.11) whilst eqn (2.8) replaced eqn (2.5) at those nodes at which eqn (2.6) was violated. The modified set of equations was then solved and inspected, this process continuing until the boundary conditions and the subsidiary conditions were satisfied at each node. Having determined the values of the forces at each surface node, the finite element method was then used to determine the state of stress throughout the components.

As a further complication, slip against finite friction on the interface results in energy dissipation so that the unloading and loading paths differ, the current state of stress depending on the loading history. In these circumstances, an incremental solution is necessary. Starting from the unloaded state, the normal load was applied in increments to the desired value, followed by application of the bending moment in increments around a cycle $0 \rightarrow M^* \rightarrow 0 \rightarrow -M^* \rightarrow 0 \rightarrow M^*$ etc. until the state of stress, the locations of the slip and separation regions, the slip and separation amplitudes reached steady-state values at successive positive peaks $M = +M^*$. Assuming that the solution corresponding to the state (N, M) has been obtained, the boundary conditions in incremental form (for ΔP_{ij}, ΔT_{ij}, $\Delta\Delta_i$) were solved for the state $(N, M + \Delta M, \Delta M$ positive or negative) by the iterative method of the previous paragraph, keeping account of the previous history of slip and separation amplitudes. The size of the load increment was chosen, with practice, to correspond to a change by one node spacing in the boundary of a slip or separation region. It was found that shakedown of the solution was usually established after $1\frac{1}{4}$ cycles of load.

2.4.1 Push–Pull Loading

In this case, only one quarter of the system of Fig. 2.7 need be considered since the plane $y = 0$ as well as the plane $x = 0$ remains plane and conditions are identical on each interface. The suffix $i = 1, 2$ in eqns (2.4) to (2.13) is now superfluous. The finite element method was used to calculate the surface displacements U_j, V_j ($j = 1, 2 \ldots n$) of the interface nodes on the plate due to a unit tensile force applied by a constant stress $\sigma_x = 1/h_1$ on the section $x = \pm b$. Incremental loading and iteration of the solution were required to obtain the regions of separation and slip when the tensile load was oscillated through the range $\pm T^*$.

2.4.2 Torsional Loading

Lau[12] studied axially-symmetric versions of the configurations of Fig. 2.5. Using axes (z, r) instead of (x, y) in Fig. 2.7, a torque T is transmitted between the free ends $z = \pm b$ of a circular shaft through a shrink-fitted assembly. The outer radius of the shaft at the interface is larger than the inner radius of the hub by an amount Δ in the unloaded state.

In such an assembly, separation of the contract surfaces under torsional loads is less likely than under bending or tensile loads. If the deformation is entirely elastic or if the coefficient of friction is zero, the circumferential torsional displacements v should be entirely uncoupled from the shrinkage displacements u radially and w axially.[14] Finite friction, however, does couple the two systems together if the surfaces slip. In a slip region, the shear stress transmitted across the interface by the torque, $\tau_{r\theta}$ is related to the shear stress set up by the shrink fit, τ_{rz}, through the law of limiting friction

$$\tau_{rz}^2 + \tau_{r\theta}^2 \leq (\mu p)^2 \tag{2.14}$$

which replaces eqns (2.6) and (2.8) above in the analysis. Furthermore, the slip displacements w_s axially and v_s circumferentially are not independent of each other. By analogy with the theorem on the increments of plastic strain in an ideal elasto-plastic material, it might be expected that the increments in the slip displacements δv_s, δw_s, caused by a change in value of the torque δT should have a resultant parallel to the direction of the resultant shear stress vector $\tau_{r\theta}$, τ_{rz} corresponding to the previous value of the torque. This condition was expressed as

$$\frac{\delta v_s}{\tau_{r\theta}} = \frac{\delta w_s}{\tau_{rz}} \tag{2.15}$$

Lau[12] found support for eqn (2.15) from a simple experiment. Equation

(2.15) together with the condition that direction of the shear stress opposes the direction of slip, replaced eqn (2.9) in the analysis.

The influence functions for the interface nodes were calculated using an axially-symmetric version of the finite element method developed by Clough and Rashid.[15] The shrink-fit deformations and the torsional loads were applied in increments and, for each increment, the solution was obtained iteratively.

2.5 STRESS ANALYSIS—RESULTS

The results of the calculations are expressed most conveniently in terms of non-dimensional load parameters, ratios of the maximum stress at sections remote from the interface to the average pressure p_0 on the interface. For the tension and bending problems, $p_0 = N/2a$; for the torsion problem, we take it to be the Lamé pressure $p_0 = E[(n^2 - 1)/n^2](\Delta/D)$ where D is the diameter of the shaft under the interface, nD is the outer diameter of the hub, Δ the diametral interference and E the Young's modulus.

At sections remote from the interface ($x = \pm b$, Fig. 2.5), the stress distributions were assumed to be those given by elementary theory:

Tension $\sigma_0 = T/2h_2$

Bending $\sigma = -\dfrac{y}{h_2}\sigma_0$ where $\sigma_0 = 3M/h_2^2$

Torsion $\tau = \tau_0 \cdot \dfrac{r}{R_2}$ where $\tau_0 = \dfrac{2}{\pi} \cdot \dfrac{M_t}{R_2^3}$

where h_2 and R_2 are the semi-thickness of the plate and the radius of the shaft at the remote sections.

We then use loading parameters, $\xi = \sigma_0/p_0$ for tension, $\lambda = \sigma_0/p_0$ for bending and $\phi = \tau_0/p_0$ for torsion. ξ^*, λ^* and ϕ^* indicate the amplitudes of these parameters under alternating load conditions. When plotting stress distributions, it will be convenient to plot stress components as fractions of σ_0, τ_0 or p_0.

Only a small selection of the results can be described here; a fuller description may be found in the theses of Wright,[8] Bramhall[11] and Lau.[12]

2.5.1 Separation of the Contact Surfaces

The form of the results depended critically on the tendency of the contact surfaces to separate when the load in the plate or shaft was applied. In the torsion problem, no such separation was found; the shrink-fit pressure distribution was quite insensitive to the value of the applied torque.

At the other extreme, considerable separation was found in the bending problem, Fig. 2.6 giving a not unrealistic view of the situation when the bending moment is at its peak value for $\lambda^* > 20$. The tendency of the surfaces to separate was found to be virtually independent of the value of the coefficient of friction.

Under tensile loading, slight separation of the contact surfaces was found for $\xi^* > 1$ and for sufficiently large μ. Separation occurs only during the tensile part of the load cycle when the frictional tractions applied to the surface of the clamping block are directed away from the centre of the interface. These tractions cause the surface of the clamping block near the edge of the interface to curl away from the interface, leading to local loss of contact. Sufficient friction is therefore necessary. While separation was found for $\xi^* > 1$ and $\mu = 0{\cdot}8$, it was absent when the calculations were repeated for $\mu = 0{\cdot}2$. During the compressive part of the load cycle, the frictional tractions on the block are reversed and the block tends to dig into the plate.

2.5.2 Stress Fields—Tension

The results of the push–pull calculations will be dealt with first. The distribution of normal pressure on the interface of the sharp assembly is shown in Fig. 2.8 at both tensile and compressive peaks of the loading cycle for $\mu = 0{\cdot}8$. Under tension, Fig. 2.8a, the normal pressure falls to zero just inside the edge of the interface for values of $\xi^* > 1$, reflecting the separation of the contact surfaces there. Contact is maintained for $\xi^* \leq 1$ and the pressure at the edge of the interface is now finite throughout the cycle. In contrast, the tendency of the surfaces to dig into each other during the compressive part of the cycle is reflected in increased pressure and a sharp stress concentration, Fig. 2.8b. On both graphs, the distribution under the action of the clamping force acting alone ($\xi^* = 0$) is included. The surfaces at the edge of the interface were found to separate by an amount $0{\cdot}2\,k$, $0{\cdot}6\,k$ and $0{\cdot}4\,k$ for $\xi^* = 2, 4$ and 8, where $k = a\sigma_0/E$. Even though separation was confined to, at most, the outer 4% of the interface, its effect on the distribution of the contact stresses is substantial.

The distribution of the axial stress σ_x at the surface of the plate is shown in Fig. 2.9 at both tensile and compressive loading peaks as a fraction of σ_0. Again, the effect of separation is evident. Whereas at the peak compressive load, Fig. 2.9b, there is a marked concentration of compressive axial stress in the plate at the edge of the interface, the tensile stress concentration is

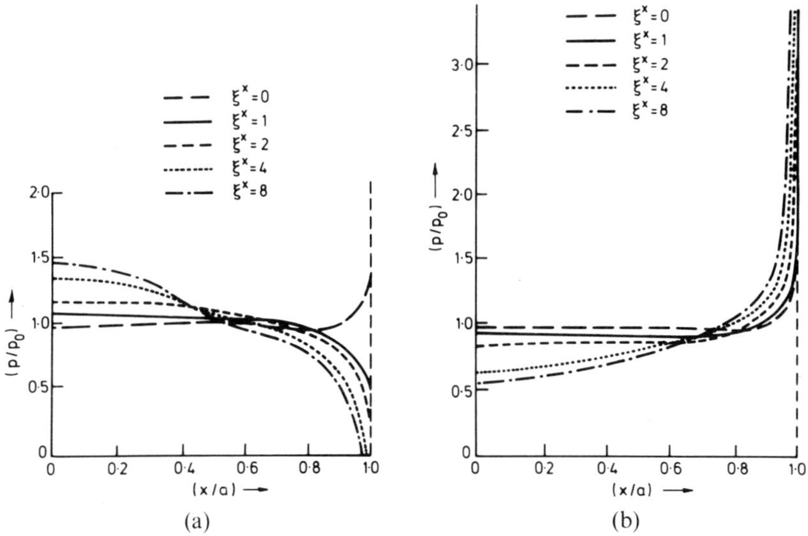

FIG. 2.8. Push–pull loading. Distribution of normal pressure in the sharp assembly (a) at the tensile peak load, (b) at the compressive peak load. Surfaces are out of contact where the pressure is zero.

somewhat less sharp. The peak tensile stress occurs at the edge of the actual contact area, at the inner boundary of the separation region. The value of the peak tensile stress there varies from about $2\sigma_0$ for $\xi^* = 1$ to $1{\cdot}07\,\sigma_0$ for $\xi^* = 8$. Separation results in almost complete collapse of the tensile stress concentration.

Figure 2.9a also shows the extent of the slip region at various values of ξ^* and $\mu = 0{\cdot}8$. The range of values of ξ^* covered the possibilities of almost complete slip ($\xi^* = 8$) and very limited slip ($\xi^* = 1$). The maximum slip amplitude at the edge of the actual contact area was found to be approximately $0{\cdot}2k, 0{\cdot}4k$ and $0{\cdot}7k$ for $\xi^* = 2, 4$, and 8, very similar in magnitude to the separation amplitude (Fig. 2.10).

The results of fatigue tests on assemblies of the same configuration will be described later but it is appropriate here to discuss the possible contribution of the stress fields just described to the initiation and growth of fatigue cracks. At large clamping pressures ($\xi^* = 1$) and with sufficient friction ($\mu = 0{\cdot}8$), the state of stress within the assembly was found to be virtually identical to a continuous component of the same shape made of one piece of material and subjected to the same loads. It might then be expected that the fatigue strength and life of such an assembly under large clamping pressures

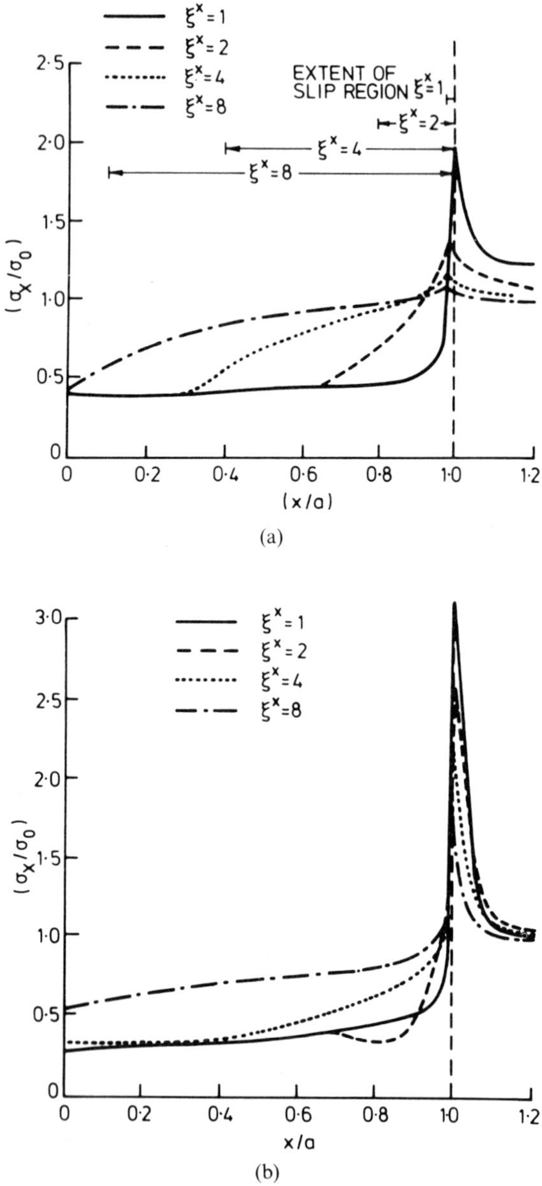

FIG. 2.9. Push-pull loading. Axial stress at the surface of the plate in the sharp assembly (a) at the tensile peak load, (b) at the compressive peak load. The position of the stress concentration in (a) coincides with the edge of the actual contact area.

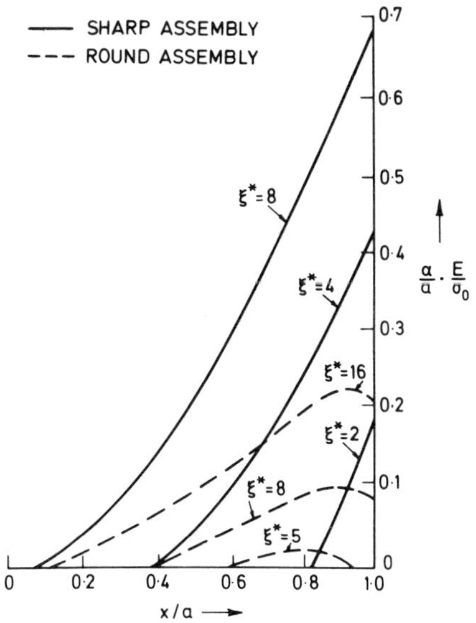

FIG. 2.10. Push–pull loading. Amplitude of slip $\alpha(x)$ along the interface in both assemblies.

would be very similar to that of the continuous component and would not be influenced by fretting.

On the other hand, at light clamping pressures ($\xi^* = 8$; $\mu = 0{\cdot}8$) the stress field in the sharp assembly differed very considerably from that in the continuous component. Anticipating the observation that the failure crack is likely to propagate from the point just inside the edge of the interface at which separation ceases and σ_x has a perceptible peak, the state of stress there was found to be:

At the peak tensile load in the plate, $\sigma_x = 1{\cdot}07\sigma_0$, $\sigma_y = 0$, $\tau_{xy} = 0$.
At the peak compressive load in the plate, $\sigma_x = -3\sigma_0$, $\sigma_y = -0{\cdot}2\sigma_0$, $\tau_{xy} = 0{\cdot}15p_0$.

Thus the plane normal to the x-axis is a principal stress plane at the tensile peak and virtually so at the compressive peak. On the plane of the likely failure crack, the normal stress therefore oscillates about a mean compressive value of $-\sigma_0$ with an amplitude of $\pm 2\sigma_0$. While these figures might suggest that the fatigue strength of the assembly at $\xi^* = 8$ would not be strikingly different from that of a plain unnotched bar of the same dimensions as the plate component, it must be remembered that the surfaces at the edge of the actual contact area slip back and forth on each other as the load oscillates and that they would therefore be liable to fret.

For all cases examined ($1 \leq \xi^* \leq 16$), no separation was found in the round assembly and the surfaces did not slip for $\xi^* \leq 4, \mu = 0{\cdot}8$. For $\xi^* \leq 4$, the distribution of stress throughout the assembly was identical to that in a geometrically similar continuous component subjected to the same loads. This result indicates how details of the geometric design, coupled with sufficiently high clamping loads, can help to eliminate slip and fretting as well as giving smoother diffusion of load into the clamped joint.

Above $\xi^* = 4$, $\mu = 0{\cdot}8$, the contact surfaces of the round assembly do slip, the size of the slip region and the amplitude of slip is shown in Fig. 2.10. With $\xi^* = 5$, the slip region is trapped within the interface with no slip at the edge. At higher values of ξ^*, slip breaks through to the edge but the maximum amplitude occurs within the interface.

In the absence of separation, the stress distributions within the round assembly had much the same form at the tensile and compressive loading peaks. For $\xi^* = 8$, $\mu = 0{\cdot}8$, the largest axial stress, $1{\cdot}25\sigma_0$, occurred at the surface of the curved transition section outside the interface. The maximum axial stress on the surface under the interface was found at $x = \pm 0{\cdot}56a$ with a value of $0{\cdot}44\sigma_0$ and the slip amplitude there was about $0{\cdot}02k$ (Fig. 2.10). It would be expected, therefore, that fatigue failure of this assembly would always occur through the curved transition section unless the contact surfaces were sufficiently weakened by fretting.

2.5.3 Stress Fields—Bending

The distribution of stress within the joint was found to be dominated by two effects, separation of the contact surfaces and diffusion of the load from the relatively thin beam outside the joint into the relatively thick beam containing the interfaces.

The onset of separation was found to differ markedly on the nominally tensile and compressive interfaces. In Fig. 2.11, the distribution of normal pressure, expressed as a fraction of the average clamping pressure, is shown at the instant of peak bending moment for a number of values of λ^*, for the sharp assembly. In Fig. 2.11, the separation regions are those on which the normal pressure is zero. On the tensile interface, separation begins at the outer edge of the interface, the extent of the separation region increasing progressively with λ^*. On the compressive interface, separation does not occur for values of $\lambda^* < 16$ and then commences inside the interface at $x \simeq 0{\cdot}7a$, the separation region spreading rapidly toward the centre $x = 0$ with further increase of λ^*. For values of λ^* in excess of about 20, contact is

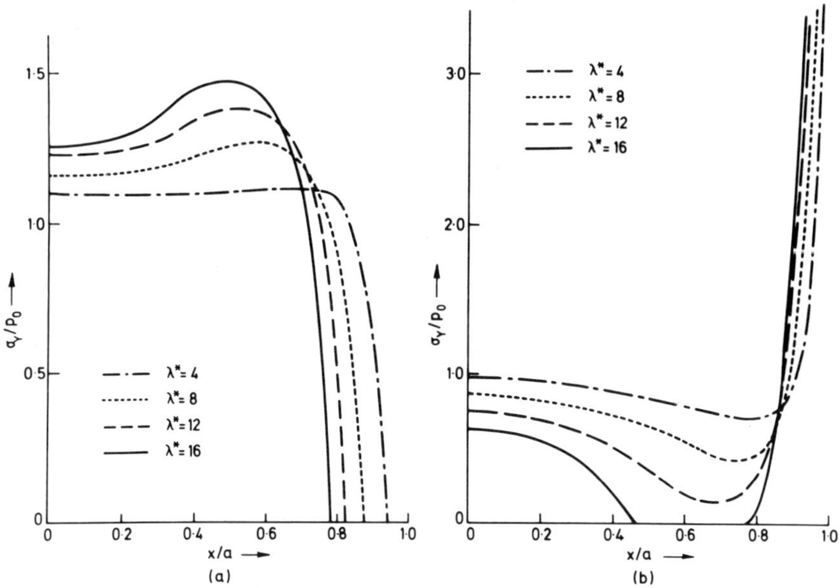

FIG. 2.11. Bending loads. Distribution of normal pressure in the sharp assembly (a) on the tensile interface, (b) on the compressive interface. Pressure is zero where the surfaces are out of contact.

lost at the centre and is then confined to a very narrow strip at the edge of the interface.

The onset of slip in the two assemblies was also found to be very similar. The tensile interface is divided into three regions, the separation region at the outside, a slip region whose outer boundary is the inner boundary of the separation region, and a no-slip region at the middle of the interface. Larger values of λ^* and smaller values of μ lead to penetration of the slip–no-slip boundary toward the centre of the interface but the actual size of the slip region diminishes at large values of λ^* when the rate of increase of the separation region exceeds that of the slip region. On the tensile interface, the slip amplitude increases from zero at the slip–no-slip boundary to a maximum value at the slip-separation boundary, the maximum slip amplitude being about $0{\cdot}1k$ for all values of λ^* at $\mu = 0{\cdot}8$. The maximum slip amplitude is relatively insensitive to the value of λ^* because the length of the separation region grows more rapidly with λ^* than the length of the slip region.

On the compressive interface, conditions are yet more complicated. For $\lambda^* \leq 16$, the slip region penetrates in from the outer edge of the interface.

For $\lambda^* \leq 4$, the surface of the plate slips inwards relative to the block. For $\lambda^* > 4$ however, the surface of the plate slips outwards relative to the block, contrary to what might be expected. This reversal in the direction of slip is thought to be due to the very different distributions of pressure on the tensile and compressive interfaces. For $\lambda^* > 20$, the clamping load is transmitted from the block to the plate through a very narrow strip of contact at the edge of the compressive interface whereas it is transmitted through the centre of the tensile interface. The contact pressures would tend to bend the plate between the interfaces in a direction opposite to that of the free plate outside the interface. For large values of λ^*, the entire contact strip at the edge of the interface slips as a peak bending moment is approached.

Although the size of the slip regions and the amplitude of slip are generally increased as the coefficient of friction falls, the pattern of separation just described is virtually independent of the value of μ.

The size of the slip and separation regions in the round assembly were much the same as in the sharp assembly at the same loads but the magnitude of the slip and separation displacements were about five times smaller.

The distribution of the axial stress σ_x on the surface of the plate in the sharp assembly is shown in Fig. 2.12 for the instant of peak bending moment. There is virtually no stress concentration peak on the tensile interface and a very sharp peak on the compressive interface, reflecting the pattern of separation within the joint. The very rapid decay of σ_x towards the centre of the joint on both interfaces indicates rapid diffusion of the bending load into the clamping blocks.

The distribution of σ_x on the tensile interface of the plate in the round assembly is shown in Fig. 2.13. The axial stress σ_x is zero at the edge of the interface where the curved surface of the plate is normal to the x-direction. Although a sharp peak is found, the magnitude of σ_x at the peak is only about $0{\cdot}15\sigma_0$. Very similar distributions occur on the compressive interface but the peak values are even smaller, about $0{\cdot}075\sigma_0$. The largest surface stress occurs in the curved transition region with a value of about $1{\cdot}06\sigma_0$.

Because of the very rapid decay of σ_x towards the centre of the joint in both assemblies, it would seem most unlikely that fatigue cracks could propagate from the contact surfaces other than from the very edge, even though the pattern of slip on the interfaces would suggest that fretting would occur. In the round assembly, the failure crack would be expected through the transition section, the presence of the joint making little difference to the fatigue strength of the plate.

For the sharp assembly, the maximum slip amplitude and the largest

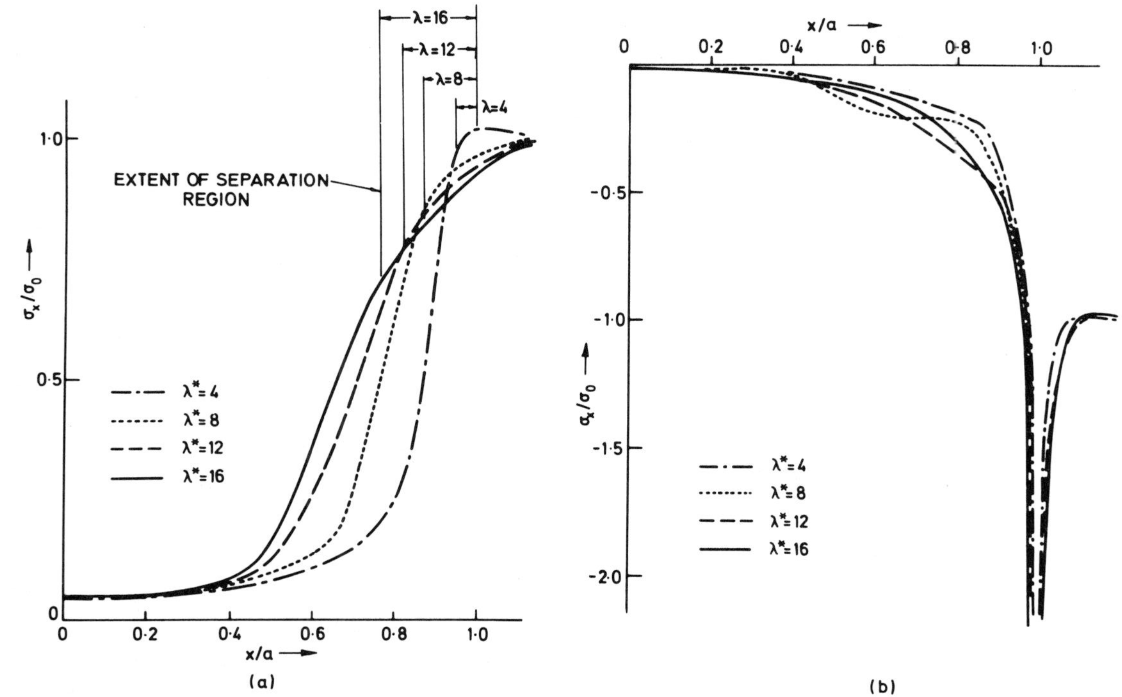

FIG. 2.12. Bending loads. Axial stress on the surface of the plate in the sharp assembly (a) on the tensile interface, (b) on the compressive interface.

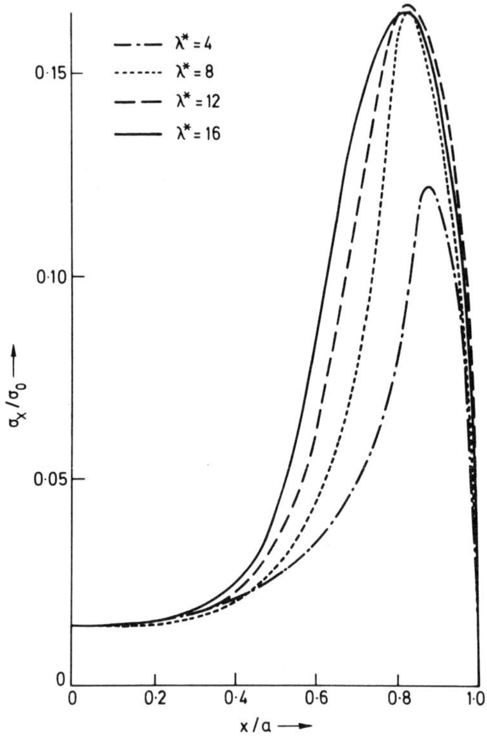

FIG. 2.13. Bending loads. Axial stress on the tensile interface of the plate in the round assembly.

stress concentration, albeit in compression, are found at the edge of the interface. Anticipating that this may be the site of a failure crack, we shall review the history of stress, slip and separation at this site as the bending load is taken through a complete cycle $M = \pm M^*$, with $\lambda^* = 4$ and $\mu = 0{\cdot}8$.

Consider the surface which is in tension when $M = +M^*$, in compression when $M = -M^*$. Starting at $M = +M^*$, the surfaces at the edge of the interface are out of contact and $\sigma_x \simeq \sigma_0$. As M is reduced from M^*, σ_x decreases more or less in proportion until contact is regained at $M \simeq +0{\cdot}5M^*$. As M is further decreased, σ_x decreases more rapidly, reaching zero when $M = +0{\cdot}2M^*$. For negative values of M, a severe compressive stress concentration occurs, with σ_x reaching a value approximately $-3\sigma_0$ when $M = -M^*$. As M is increased again, σ_x remains compressive until M reaches the value of about $0{\cdot}3M^*$. Contact is lost again when $M \simeq 0{\cdot}7M^*$ at which point $\sigma_x \simeq 0{\cdot}5\sigma_0$. As M increases to M^*, σ_x grows to the value of σ_0 again. Slip occurs throughout the proportion of the cycle during which the surfaces are in contact but it must be emphasised that, whilst the surfaces are slipping. σ_x is mostly compressive.

The exact point in the loading cycle at which contact is lost and regained depends on the values of λ^* and μ but the cycle just described and in particular, the final statement, is qualitatively typical of all values. For all values of λ^*, the maximum shear stress in the (x, y) plane reaches a sharp peak of $1{\cdot}65\sigma_0 \pm 0{\cdot}05\sigma_0$ at the edge of the interface when $M = -M^*$. Thus, at $M = \pm M^*$, the plane normal to the axis is virtually a principal plane on which σ_x varies through the range $-\sigma_0 \pm 2\sigma_0$. This conclusion is remarkably similar to that drawn from the analysis of push–pull loading.

2.5.4 Stress Fields—Torsion

These results are somewhat easier to describe because the coupling between the shrinkage stresses and the torsional stresses implied by eqns (2.14) and (2.15) proved to be too weak to result in separation of the contact surfaces. The distribution of normal pressure along the interface suffered only very slight changes when the torsional loads were applied.

White and Humpherson[16] have used a similar finite element analysis to determine the stress fields due to the shrinkage process alone but with full axial restraint ($\mu \to \infty$) or no axial restraint ($\mu = 0$). In the present work, it was found that slip of the contact surfaces in the axial direction occurs during the shrinkage process only for $\mu \leq 0{\cdot}2$ for both the sharp and round assemblies. The distribution of normal stress had the general form found by White and Humpherson, decaying from a peak value at the edge of the interface towards the Lamé value in the middle of the interface. Figure 2.14 shows the variation in the peak pressure at the edge of the interface, as a fraction of the Lamé pressure p_0, for both assemblies, plotted against μ. The relatively constant values for $\mu > 0{\cdot}2$ indicates an absence of axial slip.

Under torsional loading, it was found that the form of the results depended almost solely on the parameter $\phi^*/\mu(=\tau_0/\mu p_0)$. The torsion problem, unlike the tension and bending problems, yields self-similar solutions in the sense that the distribution of stress, the extent of the slip region, the distribution of slip along the interface, the frictional damping capacity are almost uniquely defined by the value of ϕ^*/μ. Figure 2.15 shows the extent of the slip region as a proportion of the total length of the interface plotted against ϕ^*/μ from a series of calculations in which two of the quantities p_0, τ_0, and μ were held constant while the third was varied over a wide range. If the solutions were entirely self similar with respect to

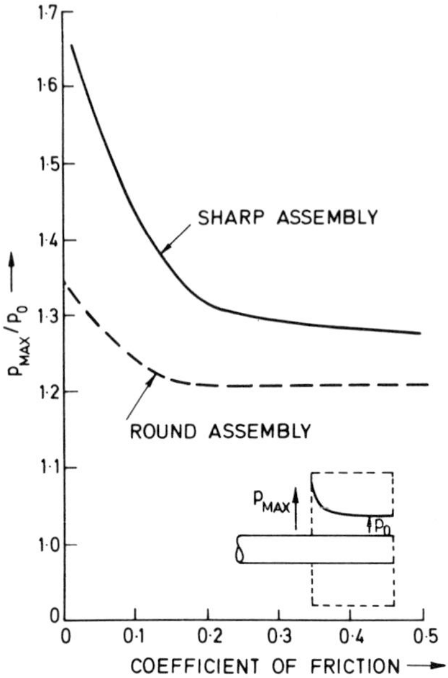

FIG. 2.14. Peak pressure at the edge of the shrink-fitted assemblies as a fraction of the Lamé pressure.

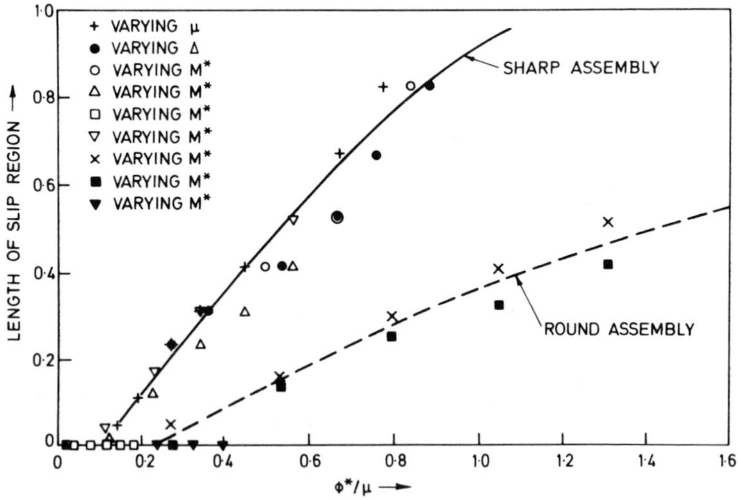

FIG. 2.15. Torsional loads. Extent of the slip region in both assemblies.

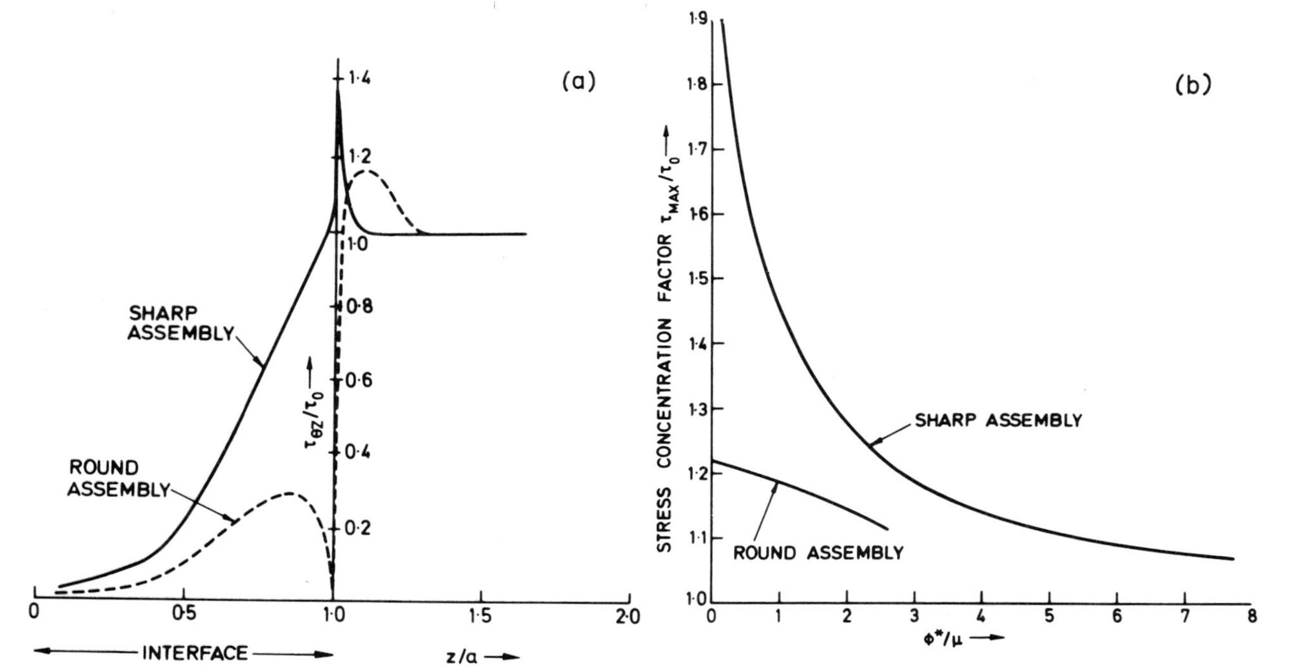

FIG. 2.16. Torsional loads. (a) Distribution of $\tau_{\theta x}$ on the surface of the shaft in both assemblies. (b) Stress concentration factors as functions of ϕ^*/μ.

ϕ^*/μ, all the points for an assembly of a given geometry should fall on a single line. The scatter in the results arises, at least in part, from the approximations inherent in the finite element method, the surface displacements being calculated only at discrete nodes on the interface. The scatter is greatest when the slip region has spread toward the centre of the interface where the nodal spacing was seven times coarser than at the edge.

It is notable that, for a given system of loads, the round assembly is less prone to slip than the sharp. By analogy with Mindlin's problem, it is unlikely that slip could be eliminated entirely in the sharp assembly for sufficiently small ϕ^*/μ, as suggested by Fig. 2.15, whereas the curtailment if not elimination of slip at high clamping pressures in the round assembly seems more plausible.

The distribution of the torsional shear stress $\tau_{\theta z}$ along the surface of the shaft in both assemblies is shown in Fig. 2.16a. In both cases, the shear stress decays rapidly towards the middle of the interface as the torque is transmitted to the hub. The sharp assembly exhibits a sharply peaked stress concentration at the edge of the interface, the more rounded maximum in the round assembly occurring at the base of the transition section. The general form of the stress distribution in both assemblies was similar at different values of ϕ^*/μ. Figure 2.16b shows the variation with ϕ^*/μ of the magnitude of the peak value of the shear stress in each assembly, expressed as a proportion of τ_0, the nominal maximum value. Large values of ϕ^*/μ correspond to large slip regions so that these results bear out the assertion made in the Introduction that slip can relieve the geometric stress concentration. They demonstrate that the price to be paid in attempting to reduce or eliminate slip by increasing the clamping pressure can be an increased geometric stress concentration. They point to the anomaly that increasing slip, while making fatigue failure due to stress concentrations less likely, makes fatigue failure induced by fretting more likely. Similar conclusions did not emerge with such clarity from the tension or bending analyses because of the complications introduced by separation of the contact surfaces.

2.5.5 Damping Capacity

Slip of the contact surfaces against finite friction gives to the assembly a frictional damping capacity. A measure of the damping capacity is usually taken to be the fractional energy dissipation $\Delta E/E$ where, for the torsion problem,

$$\Delta E = \oint M_t \, \mathrm{d}\theta$$

is the energy dissipated in a complete loading cycle and

$$E = \int_{-M_t^*}^{M_t^*} M_t \cdot d\theta$$

is the total energy input as the load swings from its peak value in one direction to its peak value in the other. ΔE is proportional to the area of the hysteresis loop formed when the torque is plotted against the angle of twist θ of a section of the shaft outside the interface. The damping capacity of both assemblies, expressed in these terms, is plotted against ϕ^*/μ in Fig. 2.17a. For each assembly, the damping capacity falls to zero at both small and large values of ϕ^*/μ. At large values there is extensive slip but the clamping

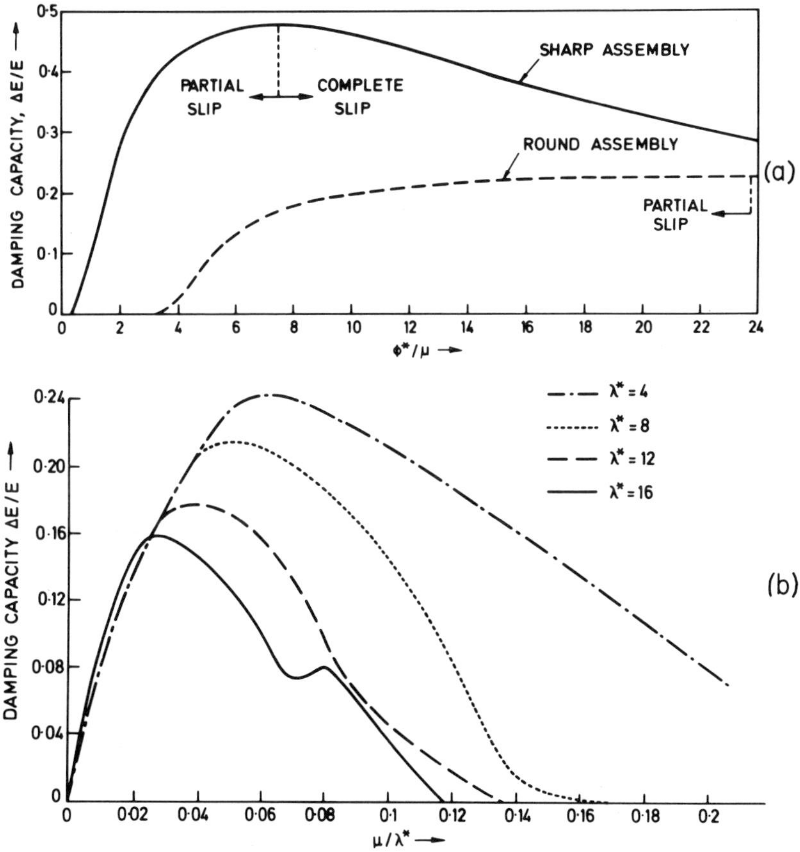

FIG. 2.17. Damping capacity (a) for the torsional loads (b) for bending loads.

pressure and hence the frictional shear stresses and energy dissipation are small. At small values, the large clamping pressure reduces to negligible proportions the size of the slip region and the amplitude of slip and hence the energy dissipation. It will be noted that a single curve is obtained for each geometry, reflecting the self-similar nature of the solutions.

In contrast, $\Delta E/E$ is plotted against μ/λ^* in Fig. 2.17b for the sharp assembly loaded in bending. Here, for a given geometry, a family of curves is obtained having λ^* as a parameter. The family of curves reflects the non-self-similar nature of those solutions, the extent of the contact areas between the components varying in a complicated way with λ^*, as already described.

Although the results of Fig. 2.17 may seem somewhat irrelevant to those endeavouring to understand or prevent the fatigue failure of clamped joints, they may be of interest to those who exploit frictional damping to control vibration (and hence fatigue failure!). Designers of turbomachinery use rings threaded through holes in the turbine blades in low-stress regions near the periphery to induce frictional damping. The presence of a maximum in each of the curves in Fig. 2.17 indicates that, in some circumstances, the design of such sacrificial contacts can be optimised.

2.6 EXPERIMENTAL

The calculations described in the previous sections were carried out as an adjunct to a programme of fatigue testing of assemblies of the geometries of Fig. 2.5. The configurations were chosen not for ease of calculation but for convenience in the laboratory. It was desirable to be able to control both the clamping and fatiguing loads.

Wright[8] and Wright and O'Connor[10] have described the push–pull testing of the configurations of Fig. 2.5 at constant values of ξ^*. Other than saying that the material used for the tests was a B.S.L65 aluminium/4% copper alloy, we will refer the reader to those sources for a description of the methods of testing. Bramhall[11] described the fatigue testing of the same configurations in the same material under alternating pure bending. He used a simple four-point-loading system in an Amsler Vibraphore to induce the alternating bending loads and clamping frames similar to those described by Wright and O'Connor[10] to apply the clamping loads.

We have not performed fatigue tests on the shrink-fitted assemblies loaded in torsion because of difficulties in manufacturing a sufficient number of specimens with a controlled range of interference fits.

2.6.1 Push–Pull Tests

The results of the push–pull fatigue tests on the sharp configurations are shown in Fig. 2.18, together with a control curve obtained from push–pull tests on a plain unnotched specimen. Results are presented for tests on the assembly at $\xi^* = 1$, 4 and 8 and for tests on solid specimens of the same geometry made from one piece of material loaded at $\xi^* = 1$ and 8.

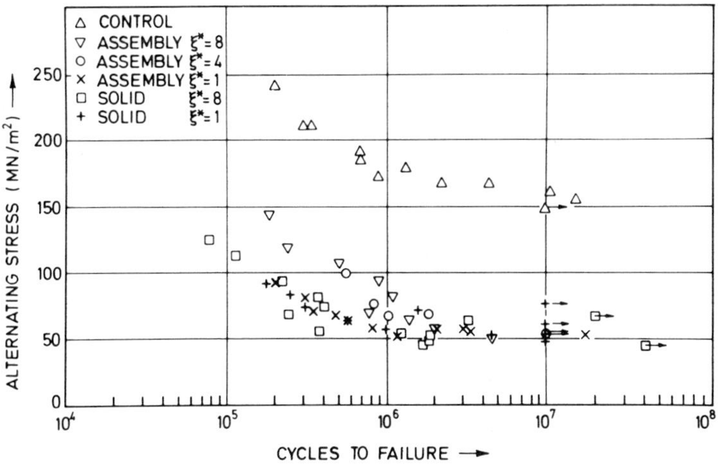

FIG. 2.18. Push–pull loads. Results of fatigue tests for the sharp assemblies and for geometrically similar solid specimens.

The results need careful interpretation. The assemblies tested at $\xi^* = 1$ yielded *S*/*N* curves very similar to those of the solid components. When compared to the control curve, there is a nominal strength reduction factor of about 3, uniform over the range of lives. All specimens failed at the sharp corner. It will be recalled that, at $\xi^* = 1$, separation of the contact surfaces was not predicted by the stress analysis, that alternating slip was confined to a very narrow band at the edge of the interface. In the experiments, it was found that fretting was also confined to a similarly narrow band of roughly the dimensions of the calculated slip band. At $\xi^* = 1$, the distribution of stress was very similar throughout the loading cycle in solid and assembled components. The fatigue tests at $\xi^* = 1$ prompt the conclusion that the strength of an assembly at sufficiently high clamping pressure is very similar to that of a geometrically similar solid component and is determined by the stress concentrations associated with the geometry.

Quite a different pattern emerged from the tests at higher values of ξ^*, lower clamping pressures. The solid components at $\xi^* = 8$ proved to be, if anything, weaker than at $\xi^* = 1$. The assembled components at short lives were found to be stronger than the solid but the strength reduction, when compared with the control curve, increased with increasing life until it reached the value of the solid for lives in excess of 10^6 cycles. Bearing in mind that the geometric stress concentration was found to collapse to 1·07 in the sharp assembly at $\xi^* = 8$ (Fig. 2.9), the increasing strength reduction with increasing life could only be attributed to the effect of fretting on fatigue strength. It was found that the failure crack always originated just inside the nominal contact area. At the tensile peak of the loading cycle, the surfaces at the edge of the interface are not in contact, the edge of the actual contact area coinciding with the position of the local peak in the distribution of σ_x (Fig. 2.9). The site of the failure crack was found to coincide quite closely with the calculated position of the local peak. Even though thc obscrvcd strength reduction could not be explained in terms of the stress distribution, the position of the failure crack could.

The effect of repeated fretting on fatigue strength was demonstrated more dramatically by the results of tests on the round component. The S/N curves for tests at $\xi^* = 16$ are shown in Fig. 2.19a together with the control curve. This value of ξ^* gives the same ratio of nominal axial stress in the specimen under the clamping blocks to normal pressure as in the sharp assembly at $\xi^* = 8$. The results of tests on assembled and solid components are virtually indistinguishable. As might be expected, all the solid components were broken by cracks through the curved transition section. The failure cracks in the assembled components occurred at the same place for short lives but some of those which broke after 10^6 cycles cracked through the thick section under the interface, the cracks originating near the calculated position of the local maximum value of the axial stress in regions of intense fretting, Fig. 2.19b. Although the nominal strength of the assembly at long lives approached that of the plain unnotched specimens, use of the results of the stress analysis indicated that the actual strength reduction factor due to fretting was about 2·6. The shift in the site of the failure crack after about 10^6 cycles demonstrates the progressive effect of fretting on fatigue strength.

Comparison of Figs. 2.18 and 2.19a shows that geometric design can increase the nominal fatigue strength of a clamped joint but it should be noted that the strength of a plain bar will be approached only if axial stress under a fretted surface is less than the nominal unnotched fatigue strength of the material by a factor exceeding the strength reduction due to fretting. Wright and O'Connor[10] reported the results of fatigue tests on a

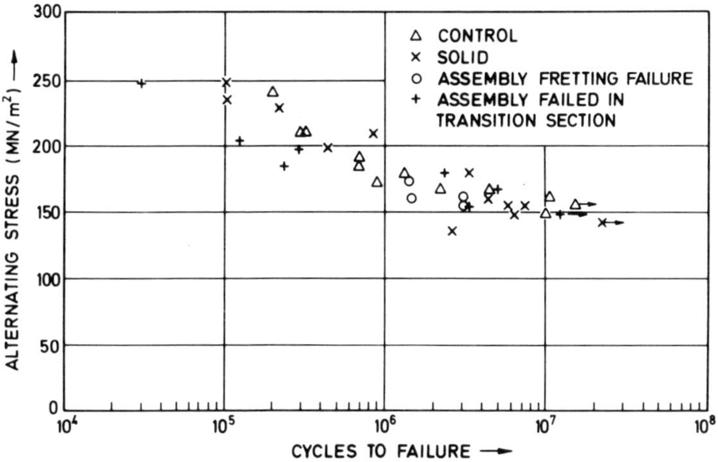

(a)

(b)

FIG. 2.19. Push–pull loads. (a) Results of fatigue tests for the round assembly and for geometrically similar solid specimens. (b) Fractured specimens. The solid specimen (top) and the assembled specimen at short lives (bottom) failed through the curved transition section. Some of the assembled specimens at long lives (middle) failed through a fretting scar on the interface.

configuration intermediate in terms of stress relief between those of Figs. 2.5a and b. Whereas the solid component proved to be slightly weaker than the plain unnotched bar, the assembly was substantially so, all but one of the failures originating at the fretted portion of the interface.

2.6.2 Bending Tests

The results of the bending fatigue tests on the sharp assembly are shown in Fig. 2.20 in which the control curve for the push–pull tests is included for comparison. Tests at different values of λ^* yielded results indistinguishable

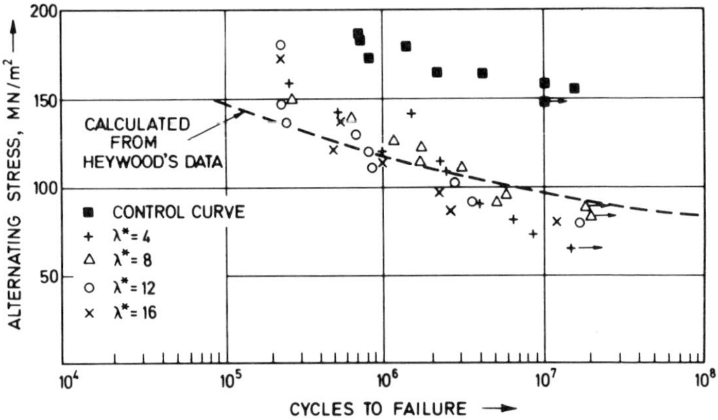

FIG. 2.20. Bending loads. Results of fatigue tests on the sharp assembly.

from each other all showing a nominal strength reduction of about 2, reasonably uniform over the range of lives. All the assemblies failed at the sharp corner. We stated above that, for all values of λ^*, the axial stress at the edge of the interface oscillated with an amplitude $\pm 2\sigma_0$ about a mean compressive value of $-\sigma_0$. While we have not found information on the effect of mean stress on the fatigue strength of B.S.L65 (UTS 510 MN m^{-2} (33 tons m^{-2})), the material used in all these experiments, Heywood[17] gives data for a similar aluminium alloy with a UTS of 528 MN m^{-2} (34·2 tons m^{-2}). Using Heywood's data and the calculated state of stress at the edge of the interface, the dotted curve in Fig. 2.20 was obtained. The dotted curve fits the results of our fatigue tests quite well, strongly suggesting that the strength of these assemblies was largely controlled by the geometric stress field.

We can state this conclusion with some confidence even though the contact surfaces were found to be severely damaged by fretting. It is worth recalling that the surfaces at the edge of the interface are out of contact during much of the tensile portion of the loading cycle, that they are in contact, slipping and fretting only when the axial stress in the specimen at the failure site is compressive (cf. Section 2.5.3). It may then be concluded that fretting applied only during the compressive portion of the load cycle does not weaken the material. By analogy with the results of the push–pull tests, it might be expected that fretting fatigue cracks would originate at the edge of the actual contact area at the tensile peak of the load cycle. No such cracks were found. Reference to Fig. 2.12 shows that the axial stress in the specimen decays very rapidly as the bending load is transmitted into the clamping blocks. Even if fretting fatigue cracks were initiated along the interface, the tensile axial stress in the specimen at such sites was clearly too small to begin the process of propagation before failure occurred at the edge of the interface.

The results of tests on the round assembly at various values of λ^* were indistinguishable from the control curve. Despite fretting on the interface, all the specimens broke in the curved transition section where the calculations predicted a modest stress concentration. No fretting fatigue failures through the interface occurred. The very rapid decay of the axial stress under the interface (Fig. 2.13) ensured that the level of stress at the site of any cracks initiated by fretting was far too low to propagate such cracks to failure. The results of these tests are therefore the clearest vindication of the use of geometric design to improve the fatigue strength of clamped joints. They point to two criteria to be satisfied by a successful design. The geometry must be such that no substantial stress concentrations arise and the level of stress at any contact surface liable to fretting damage must be less than the nominal fatigue strength of the material by a factor greater than the maximum strength reduction caused by fretting.

2.7 THE MECHANISM OF FRETTING FATIGUE

When the results of the fatigue tests just described were analysed in terms of the calculated elastic stress fields in the specimens, it was necessary in some circumstances to attribute the initiation of failure cracks to the action of fretting. The question then arises why the production of fretting damage on the surfaces of a specimen subjected to alternating stress can reduce fatigue strength so drastically.

There appears to have been no quantitative attempt to examine the assertion made by Fenner and Field[18] that the abrasive action of fretting removes adherent oxide layers, 'so permitting intermetallic welding on a scale sufficient to produce high stresses, due to tangential displacement, in a volume of material large enough for definitive cracks to be developed'. They based their assertion mainly on two observations. The strength reduction due to fretting increases with increasing life. The coefficient of friction increases with increasing life. The fate of the round assemblies in push–pull, described above, confirms the first observation. Both Wright and O'Connor[10] and Bramhall[11] found that the average coefficient of friction between the surfaces of their B.S.L65 specimens rose from an initial value of 0·2, typical of boundary lubrication, to 0·8, typical of substantial intermetallic contact. Many others have made similar observations. Wright and O'Connor reiterated the Fenner and Field assertion in a slightly modified form: 'growth in the coefficient of friction with simultaneous growth in the areas of real contact are essential features of the fretting fatigue mechanism'.

In order to examine these assertions, it is necessary to change the scale of our discussion from the macroscopic stress fields associated with the geometry and means of loading of an assembly to the microscopic stress fields associated with the regions of actual contact at the tips of the highest asperities on the contact surfaces. But how does one describe those microscopic stress fields?

O'Connor and Johnson[4] found some evidence to suggest that microscopic asperity contacts, even if formed initially by bulk plastic deformation, deform elastically after the first few cycles of alternating frictional load. They also observed that individual asperity contact strips on geometrically rough surfaces can exhibit fretting patterns very similar to those found by Johnson[3] in the contact of metal spheres, i.e. a central undamaged strip surrounded by fretted strips. There may be some justification, therefore, in discussing asperity contacts in terms of the Hertz/Mindlin/Poritsky theory.

Buckley[19] described a mechanism of crack initiation at frictional contact in which high tensile stresses at the trailing edge of the contact area cause local fracture along slip-bands lying oblique to the surface. The existence of slip bands at the surface implies local yielding there and, as shown by Hamilton and Goodman,[6] yielding at the surface requires sufficiently high friction. We also observed, Figs. 2.3 and 2.4, that substantial tensile fields also require sufficiently high friction. Thus, the observed growth in the coefficient of friction in fretting contacts would seem to be an essential precondition for crack initiation.

Presumably a 'definite crack' is one sufficiently long that it can be propagated to failure by the bulk alternating stress within the specimen. The length of a crack initiated by Buckley's mechanism at an asperity contact should be directly related to the diameter or width of that contact. There may then be a critical minimum diameter of asperity contact needed for the formation of a definite crack. If tensile stress is needed to elongate such a crack, we have seen (Figs. 2.3 and 2.4) that the volume of material subjected to tension by a frictional contact increases very rapidly with increasing coefficient of friction. The critical minimum diameter should therefore decrease with increasing coefficient of friction.

These are propositions which can be tested experimentally, by performing fatigue tests on specimens to the surfaces of which alternating frictional contact stresses are applied. To help with the interpretation of the results of such tests, it would be desirable to keep the amplitude of the contact stresses and the bulk propagating stress in the specimen constant from test to test while varying only the size of the contact area. The specimens would fail in fatigue only when the size of the contact area exceeds the critical value.

The Hertz/Mindlin/Poritsky theory provides a means for designing such experiments. If cylindrical indentors of radius R are pressed against the plane surfaces of a fatigue specimen of the same material by a force N per unit length of cylinder, the semi-width of the contact strips formed is given by

$$a = 1[1RN(1 - \nu^2)/\pi E]^{1/2} \tag{2.16}$$

and the maximum Hertzian pressure is

$$p_0 = \frac{2N}{\pi a} = \left[\frac{E}{2\pi(1 - \nu^2)}\right]^{1/2} \left[\frac{N}{R}\right]^{1/2} \tag{2.17}$$

where E is the Young's modulus and ν the Poisson's ratio of the materials. If the fatigue tests are performed with a range of indentor sizes but if the ratio N/R is kept constant from test to test, then the maximum contact pressure would also be held constant while the size of the contact areas could be varied in a controlled way from test to test. If alternating tangential forces T precisely equal to limiting friction could be transmitted to the specimen, then the maximum tensile stress induced in the specimen by the indentors, $2\mu p_0$, would also be held constant from test to test. Even if $T/N < \mu$, the contact stresses at any point in the specimen are proportional to p_0. We have, therefore, a means of applying to the fatigue specimens a given distribution of frictional contact stress of given maximum amplitude while varying the volume of material subjected to that contact stress field.

2.7.1 The Experiment

A diagram of the experimental apparatus is shown in Fig. 2.21; a full description of the apparatus and method of testing was given by Bramhall.[11] The two cylinders were pressed against the fatigue specimen by a simple loading frame. The movement of the cylinders parallel to the axis of the specimen was constrained by a pair of springs, metal cantilevers attached to the fixed frame of the testing machine. The stiffness of the springs was varied from test to test in a given series in an attempt to keep the ratio T/N constant. Eight cylindrical indentors, with radii in the range 1·27–20·32 cm ($\frac{1}{2}$–8 in), were used.

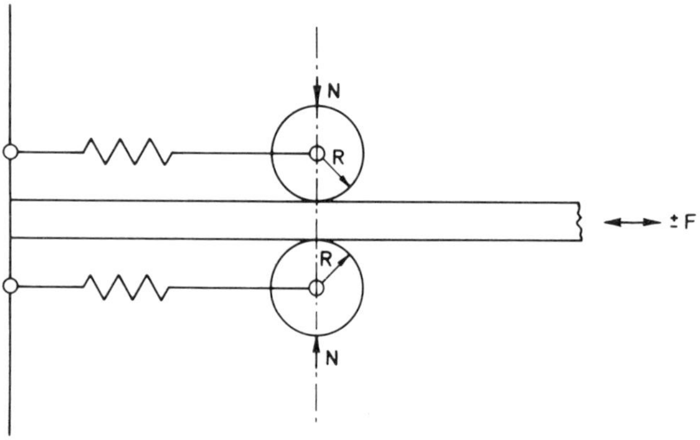

FIG. 2.21. Diagram of apparatus.

The material used for both fatigue specimens and cylindrical indentors was B.S.L65 aluminium alloy, the same material as for the push–pull and bending tests already described.

After each fatigue test, the lower clamp on the specimen was released and the static load device on the fatigue machine was used to determine the value of the average† coefficient of sliding friction between the indentors and the specimen.

† The value of μ measured after fretting had taken place on part of an interface would be some average of the high local values on the fretted part of the contact surface and the lower local values on the non-fretted part.

2.7.2 Test Conditions

Three values of N/R were used, 1·08, 1·54, 1·85 MN m^{-2}, giving values of p_0 from eqn (2.17) of 118·6, 141·8, 155·3 MN m^{-2} respectively. For each value of N/R, tests were carried out at three values of the nominal alternating stress, σ_0, in the specimen, 61·7, 77·2 and 92·7 MN m^{-2}. In these tests, the value of T/N was adjusted to give a final measured value of the average coefficient of friction of about 0·3. In two further series, with $N/R = 1{\cdot}54$ MN m^{-2}, $\sigma_0 = 108{\cdot}1$ and 123·6 MN m^{-2}, the measured value of μ_{av} was in the range 0·4–0·5. All these tests were carried out in conditions of partial slip to avoid excessive wear of the indentors and loss of control over the contact stresses. One series of tests was carried out in conditions of gross sliding, with $N/R = 1{\cdot}08$ MN m^{-2} and $\sigma_0 = 92{\cdot}7$ MN m^{-2}, giving final μ_{av} values in the range 0·8–0·9. Note that the unnotched fatigue strength of the material at 10^7 cycles is 155 MN m^{-2}.

Attempts were made to measure a'/a, the ratio of the widths of the non-fretted to contact strips, as a check on the value of T/N in each test. However, the contact areas were rarely truly straight with parallel edges, the value of a'/a varied over the width of the specimen and the slip/no-slip boundary proved difficult to locate with precision under the travelling microscope. It was apparent that, as the coefficient of friction rose, the slip region diminished so that slight fretting damage could be observed over much of the final relatively undamaged non-slip region. The variations of μ and a'/a throughout the course of each test and the variations of a and a'/a across each contact strip made precise correlation of the experimental results with the plane strain theory difficult.

2.7.3 Experimental Results

Despite the departures of the experimental system from the ideal, the results of each series of tests showed a similar pattern. For a given value of N/R and of σ_0, it was found that the specimens remained unbroken after 10^7 cycles of load until the radius of the indentors reached a certain critical value, corresponding to a critical contact width, after which a sharp reduction in life was observed. Results for some of the series of tests are shown in Fig. 2.22 and the values of the critical contact widths for the twelve series are given in Table 2.1, together with the averages of the measured values of the coefficient of friction for each series.

The values of the critical contact width were calculated from eqn (2.16), using the appropriate critical value of R. In each case, the lower value corresponds to the largest value of R at which the specimen remained unbroken after 10^7 cycles, the higher value corresponding to the next

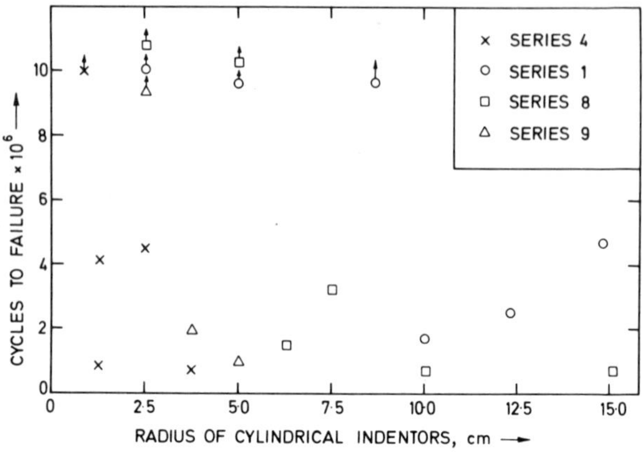

FIG. 2.22. Fatigue life plotted against indentor radius for four series of fatigue tests in each of which N/R was held constant.

highest value of R used which resulted in a failure in less than 10^7 cycles. In the gross slip series 4 tests, wear of the indentors resulted in contact strips two to three times wider than the theoretical values.

To obtain these results, about 120 fatigue tests were performed. In order to save specimens and time, specimens which were unbroken after 10^7 cycles were re-used with larger indentors located further along the test section. No subsequent failures were obtained from contact strips of less than the

TABLE 2.1
VALUES OF THE CRITICAL CONTACT WIDTHS

Series No.	$p_0\,(MN\,m^{-2})$	$\sigma_0\,(MN\,m^{-2})$	μ_{av}	*2a critical (mm)*
1	118·6	61·6	0·3	1·47–1·78
2	118·6	77·1	0·3	1·04–1·17
3	118·6	92·5	0·3	0·74–0·89
4	118·6	92·5	0·8–0·9	0·08–0·15
5	141·8	61·6	0·3	1·75–2·10
6	141·8	77·1	0·3	1·07–1·42
7	141·8	92·5	0·3	0·53–0·71
8	141·8	107·9	0·4–0·5	0·71–0·89
9	141·8	123·3	0·3–0·5	0·36–0·53
10	155·3	61·3	0·3	1·55–1·93
11	155·3	77·1	0·3	1·17–1·35
12	155·3	92·5	0·3	0·58–0·76

critical width although some of these specimens were subjected to a total of 3×10^7 cycles of fatigue load.

The experimental results indicated that the size of the contact area must exceed a certain critical value before a given fatigue stress amplitude σ_0 and (within reason) a given distribution of contact stresses can initiate propagating cracks. For a given contact stress field (N/R and μ_{av} constant), the critical size decreases as σ_0 increases. This result is to be expected since the state of stress in the vicinity of the contact depends on the superimposed effects of the contact stress field and the bulk stress.

Comparing series 3 and 4, the dramatic reduction in the critical contact size with increasing μ_{av} is consistent with the effects of friction on the tensile stress field, as discussed above (Figs. 2.3 and 2.4).

In the low average friction partial slip tests, the values of critical contact width observed probably exceed the usual notions of asperity contact size. This fact itself could be significant since it implies that a low coefficient of friction would require improbably large areas of real contact for the initiation of failure cracks. The critical contact width quoted for the high friction series 4, whilst substantially smaller, is open to doubt because of indentor wear. The results, however, appear to give at least qualitative support to the proposition that there is a minimum size of contact area associated with the initiation of propagating cracks and that the minimum contact size is strongly dependent on the value of the coefficient of friction.

The present results apply only to the alloy B.S.L65 used in the experiments. However, Waterhouse[20] exploited eqns (2.16) and (2.17) to examine the effect on fatigue life of increasing p_0 whilst keeping a constant. Using values of a equal to 25 μm and 50 μm, and performing fatigue tests on a 70/30 brass and an Al–Zn–Mg alloy, he concluded that there is a value of p_0 below which fatigue life is not seriously shortened and that the (critical) value of p_0 is smaller for the broader contact strip. Since, for a given value of μ, the maximum tensile stress is proportional to p_0, Waterhouse's results may not be inconsistent with the present. Indeed, his results for the brass specimens could be taken to demonstrate that, with $p_0 = 200\ \mathrm{MN\,m^{-2}}$, $\sigma_0 = 154 \pm 143\ \mathrm{MN\,m^{-2}}$, the series with $a = 50\ \mu\mathrm{m}$ gave a life of about 2×10^5 cycles whilst $a = 25\ \mu\mathrm{m}$ would give a life in excess of 10^7 cycles, consistent with the present results.

The quantity $\sigma_0 + 2\mu_{av}p_0$ may be taken as some measure, however imperfect, of the maximum stress in the specimen. Apart from the complete slip series 4 tests, it is a quantity which probably remained reasonably constant from test to test in any one series. Theories of fatigue crack

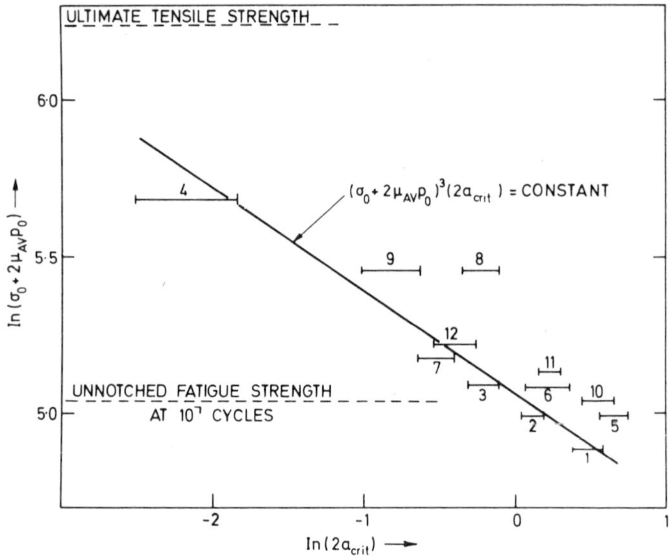

FIG. 2.23. Variation of critical contact width with $\sigma_0 + 2\mu_{av}p_0$.

propagation,[21] suggest that the value of the nominal stress σ required to propagate a crack of length l to failure is given by the relationship

$$\sigma^3 l = \text{constant} \tag{2.18}$$

Figure 2.23 shows a plot of the values of $\ln(\sigma_0 + 2\mu_{av}p_0)$ against $\ln(2a_{crit})$ for all twelve series of tests reported here. A straight line of slope 3 is included. Whilst there are obviously too few data, particularly in the high stress/low contact width range, to enable firm conclusions to be drawn, there does appear to be a pattern of behaviour similar to that of materials in more conventional fatigue situations. If this pattern could be confirmed by further experiment, it might help to bring the problem of fretting fatigue more squarely within the embrace of the theory of fracture mechanics.

2.8 CONCLUSIONS

1. Elastic stress analysis can be used to predict the incidence of interfacial slip in clamped joints. Under conditions of alternating load, the predicted slip regions compare well with observed regions of fretting damage.

2. For other than the simplest configurations, numerical methods of analysis based on finite elements can be used.

3. The state of stress within assembled components depends critically on whether the contact surfaces separate under load. In the examples described here, bending loads produced substantial separation; tension/compression loads produced modest but significant separation and torsional loads produced no separation.

4. The calculated stress fields in clamped joints facilitate the interpretation of the results of fatigue tests on similar configurations. The reduction in fatigue strength due to fretting can be calculated more accurately when the macroscopic distribution of stress within the joint is known.

5. The strength of a simple joint loaded in bending was found to be determined entirely by the macroscopic stress field, despite the incidence of fretting.

6. When the same joint was loaded in tension/compression, the fatigue strength was determined by the macroscopic stress field only at high clamping loads. At low clamping loads, fretting was the cause of failure.

7. For both methods of loading, geometric design aimed at the elimination of stress concentrations gave substantial improvement in fatigue strength. Fretting fatigue failures can be avoided only when level of stress at a fretted interface is less than the unnotched fatigue strength by a factor greater than the strength reduction due to fretting.

8. The strength reduction due to fretting increases with increasing life. It is suggested the action of fretting is to increase the areas of real contact on the interface and to increase the coefficient of friction until a sufficient volume of material is subjected to a sufficiently high contact stress field for a failure crack to be initiated.

9. Fretting fatigue tests with cylindrical contacts confirm that a minimum size of contact area is needed to initiate a failure crack and that the minimum size decreases with increasing coefficient of friction.

ACKNOWLEDGEMENTS

This chapter could not have been written without the contributions of my three former research students, Drs G. P. Wright, M. G. Lau and R. Bramhall. The work was supported by the Ministry of Defence and in its initial stages, by the Science Research Council. Mrs J. Takacs kindly traced the diagrams and Miss C. Burden kindly and patiently typed the manuscript.

REFERENCES

1. Cattaneo, C., Sul contatto di due corpi elastici: distribuzione locale degli storzi. *Pend. d'Acad. naz. dei Lincei*, **27**, Series 6, pp. 342, 434, 474 (1938).
2. Mindlin, R. D., Compliance of elastic bodies in contact. *J. Appl. Mech.* (Trans. ASME, Series E), **16**, 259 (1949).
3. Johnson, K. L., Surface interaction between elastically loaded bodies under tangential forces. *Proc. Roy. Soc. Lond. A*, **230**, 531 (1955).
4. O'Connor, J. J. and Johnson, K. L., The role of surface asperities in transmitting tangential forces between metals. *Wear*, **6**, 118 (1963).
5. Goodman, L. E. and Brown, C. B., Energy dissipation in contact friction; constant normal and cyclic tangential loading. *J. Appl. Mech.* (Trans. ASME), **84**, 17 (1962).
6. Hamilton, G. M. and Goodman, L. E., The stress field created by a circular sliding contact. *J. Appl. Mech.* (Trans. ASME Series E), **33**, 371 (1966).
7. Poritsky, H., Stresses and deflections of cylindrical bodies with application to contact of gears and locomotive wheels. *J. Appl. Mech.* (Trans. ASME, Series E), **17**, 191 (1950).
8. Wright, G. P., Studies in fretting fatigue. D.Phil.Thesis, Oxford University (1970).
9. Wright, G. P. and O'Connor, J. J., Finite-element analysis of alternating axial loading of an elastic plate pressed between two elastic rectangular blocks with finite friction. *Int. J. Engng. Sci*, **9**, 555 (1971).
10. Wright, G. P. and O'Connor, J. J., The influence of fretting and geometric stress concentrations on the fatigue strength of clamped joints. *Proc. Instn. Mech. Engrs*, **186**, 827 (1972).
11. Bramhall, R., Studies in fretting fatigue. D. Phil. Thesis Oxford University (1973).
12. Lau, M. G., Shrink-fitted assembly subjected to cyclic torque. D. Phil. Thesis, Oxford University (1972).
13. Zienkiewicz, O. C., *The finite element analysis in engineering science*. London, McGraw Hill (1971).
14. Timoshenko, S. and Goodier, J. N., *Theory of elasticity*, 2nd edn, New York, McGraw Hill p. 306 (1951).
15. Clough, R. W. and Rashid, Yusef, Finite element analysis of axisymmetric solids. *Proc. A.S.C.E.*, **91**, EM1, 71 (1965).
16. White, D. J. and Humpherson, J., Finite element analysis of stresses in shafts due to interference fit hubs. *J. Strain Anal.*, **4**, 105 (1969).
17. Heywood, R. B., *Designing against fatigue*, London, Chapman and Hall p. 45 (1962).
18. Fenner, A. J. and Field, J. E., A study of the onset of fatigue damage due to fretting. *Proc. N.E. Coast Instn Engrs and Shipbuilders*, **76**, 183 (1960).
19. Buckley, D. H., Effect of various material properties on the adhesive stage of fretting. Agard-CP-161. Paper No. 13 (1975).
20. Waterhouse, R. B., *Fretting corrosion*, London, Pergamon, p. 140 (1972).
21. Frost, N. E., Crack propagation in an aluminium alloy. *J. Mech. Engng. Sci.*, **2**, 109 (1960).

3

The Application of Fracture Mechanics to Predicting Fretting Fatigue

P. R. EDWARDS

Structures Department, Royal Aircraft Establishment, Farnborough, Hants, UK

LIST OF SYMBOLS

a	Crack length
A	Cross sectional area of specimen
e	Fretting scar width
F_f	Maximum frictional force in stress cycle on each pad foot
F_n	Normal force on each pad foot
f	Function
K	Stress intensity factor
K_a	Semi-range alternating stress intensity factor
K_d	Length correction stress intensity factor
K_m	Mean stress intensity factor
K_{fp}	Stress intensity factor due to frictional load of 1N on fretting scar of width 8 mm
K_{fp1}	As K_{fp} but without component due to reacting stresses
K_{fp2}	Component of K_{fp} due to reacting stresses
K_{IC}	Plane strain fracture toughness
K_{np}	Stress intensity factor due to normal load of 1N on fretting scar width of 8 mm
N	Number of cycles to failure
$p(K_a)$	Probability distribution of K_a
w	Distance between fretted surfaces (specimen depth)
σ_a	Semi-range alternating body stress in specimen
σ_r	rms alternating stress
σ_c	Semi-range alternating body stress in specimen to give threshold of crack propagation
μ	Coefficient of friction

3.1 INTRODUCTION

Fretting occurs where there is repeated movement between metal surfaces in contact. Its effect on fatigue is to speed up the initiation of fatigue cracks and it can be extremely damaging. For instance, it has been shown[1] for a clamped joint in which fatigue failure occurs in a plain aluminium alloy centre plate from the fretting action of steel side plates, that the alternating stress required to cause failure in 10^7 cycles was less than one third of that for a plain aluminium alloy specimen without fretting. Similarly, fretting between a close-fitting pin or bolt and the bore of a hole in, say, a lug can be very damaging.[2] It has been suggested by a number of investigators[3-7] that the damaging effect of fretting depends upon the magnitudes of the frictional forces between fretted surfaces. In particular, Endo and Goto[7] have carried out crack growth measurements on fretted specimens of mild steel. They showed that cracks initiated on shear planes (stage I) and then very quickly started propagating approximately normal to the surface (stage II) from a depth of about 0·03 mm. There followed a period of crack growth at a rate which it was shown was highly dependent on the alternating frictional force in the fretted area. After approximately 25 % of the life had been consumed, the crack had grown away from the area over which frictional forces were being applied and it was found that frictional forces no longer affected the crack rate. Crack rates calculated from tests at a range of alternating stress levels then correlated well with the alternating stress intensity factor computed, taking into account only the alternating body stress in the specimen. Since the crack rates in the early stages of the life are dependent upon frictional forces, it seems likely that the stress intensity factor in this region contains a component which also depends upon the frictional forces. Further, the stress intensity factor at short crack lengths is also likely to contain a component which depends upon the normal fretting forces. Incorporating fretting force components into the computed stress intensity factor should enable more accurate predictions of crack rate to be made over the early part of the life in which fretting forces are significant. Although the percentage of the fretted life over which fretting forces are dominant is normally small (25 % in the example above) it should be appreciated that were there no fretting forces, the number of cycles to cover the corresponding range of crack lengths would typically be multiplied many times even if the crack would propagate at all. This suggests that fretting forces, used with an associated fracture mechanics model, could be used to predict the observed, sometimes very large, changes in life which occur when fretting conditions are varied.[8]

This chapter describes the development of such a fracture mechanics model for the life prediction of fretting fatigue[9,10] and summarises the measurement of frictional forces under constant[11] and variable[12] amplitude loading. The frictional forces were required in order to scale the value of associated stress intensity factor. Also described is the application of the model to predict the lives of specimens under constant[10] and variable[13] amplitude loading. The specimens to which the model was applied were of aluminium alloy with steel fretting pads. Predictions are shown for lives obtained originally[1] under the two loading actions for four values of pad span (i.e. different values of slip) and two values of pad load. Additionally, predicted crack growth curves are compared with limited data available.[14]

This work represents a stage in the development of a fracture mechanics analysis of a range of structural components in which failure is due to fretting, with subsequent application to both conventional fatigue life estimation and damage tolerant design.

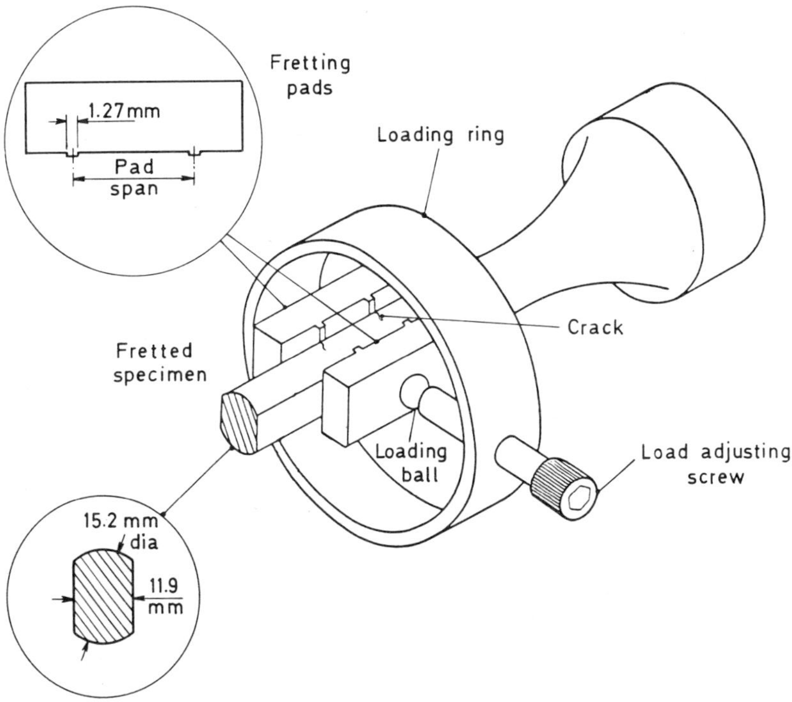

FIG. 3.1. Fretted specimen assembly.

3.2 SPECIMEN AND MATERIAL

A specimen is shown in Fig. 3.1 together with the fretting pads. The pads were clamped to the test specimen on both sides as shown, using a steel ring and balls. The clamping force was measured by strain gauges attached to the ring and also by ring deflection. The material used for the specimen was to specification BS L65, being the fully artificially-aged version of a 4% copper–aluminium alloy. Measured tensile properties of the material are given in Table 3.1. The fretting pads were made from BS S98 steel. Pad spans used were 2 mm, 6·35 mm, 16·5 mm and 34·35 mm.

TABLE 3.1
MEASURED MECHANICAL PROPERTIES OF BS L65 ALUMINIUM ALLOY USED IN THE FATIGUE TESTS (AFTER REFERENCE 1)

0·1% Proof stress	480·6 MN m^{-2}
0·2% Proof stress	488·1 MN m^{-2}
UTS	542·4 MN m^{-2}
Elongation on 50 mm gauge length	12·5%

3.3 TEST CONDITIONS

All tests were carried out at zero mean stress and at a frequency of 120 Hz in a resonant fatigue machine modified[15] to apply Gaussian narrow band random loading as well as constant amplitude loading. The random loading had the form of a randomly modulated sine wave with a nominal Rayleigh distribution of peaks. This distribution was monitored and significant variations from the Rayleigh distribution were found which depended upon the magnitude of the rms stress of the waveform. One particular type of specimen was designated as 'standard'. This was a specimen having a pad load of 2·5 kN and a pad span of 16·5 mm. This was the only type of specimen tested as a part of both investigations, i.e. into the effect of pad span and load.

Lives to failure for the tests with four pad spans and two pad loads (2·5 kN and 1·0 kN) under the two loading actions are given in Section 3.6, where they are compared with the fracture mechanics predictions.

3.4 METHOD OF CALCULATION OF STRESS INTENSITY FACTORS, CRACK RATES AND LIVES

3.4.1 Stress Intensity Factors

The stress intensity factor at the tip of a crack growing from the end of a fretting pad was assumed to be made up from three individual contributions. First was the contribution due to the body stress in the specimen[16] and which would have been the only contribution had there been no fretting pads. Second was the contribution due to the alternating frictional loads. Both these contributions were alternating and hence damaging. The third contribution was that due to normal pad loads. This contribution was static and compressive. Hence it tended to be beneficial. Only mode I stress intensity factors were taken into account. Although mode II (shear) stress intensity factors were present, no account was taken of them as no appropriate data was available on which to base predictions.

The pad load contributions to the stress intensity factor were calculated by integrating over the fretted surface the solutions of Rooke and Jones[17] for tangential and normal line loads at a varying distance from a crack of varying length. It was assumed in all cases, as was found in practice, that the cracks initiated from the outer edges of the pad feet (Fig. 3.1). It was also assumed that the cracks grew normal to the surface. This was not quite representative because in practice it was found that in the initial stage II crack growth the cracks slanted slightly so as to grow under the pads. Since it was found that straight-through cracks were common in these specimens it was assumed that the distributions of pad loads were uniform across the widths of the fretting pads. However, no such assumption could be made about the distributions of loads over the 1·27 mm between the 'front' and 'back' of the fretting pad feet. Stress intensity factors were therefore evaluated for a range of assumed distributions of pad load as shown in Fig. 3.2, which included the extreme cases of all the loads concentrated either at the front or back of the fretting pad foot. It was assumed that the actual stress intensity factor involved a distribution of pad load somewhere between these two extreme situations.

Figures 3.3 and 3.4 show the stress intensity factors K_{fp} and K_{np} for the range of assumed pad load distributions, due to frictional and normal loads respectively assuming in each case a load of 1 N and a fretting scar width (the longer dimension of the fretting scar) of 8 mm. As can be seen from Fig. 3.3, the stress intensity factor due to frictional forces is split into two components, K_{fp1} which varies with the assumed pad load distribution, and K_{fp2} which is negative and accounts for the fact that the specimen

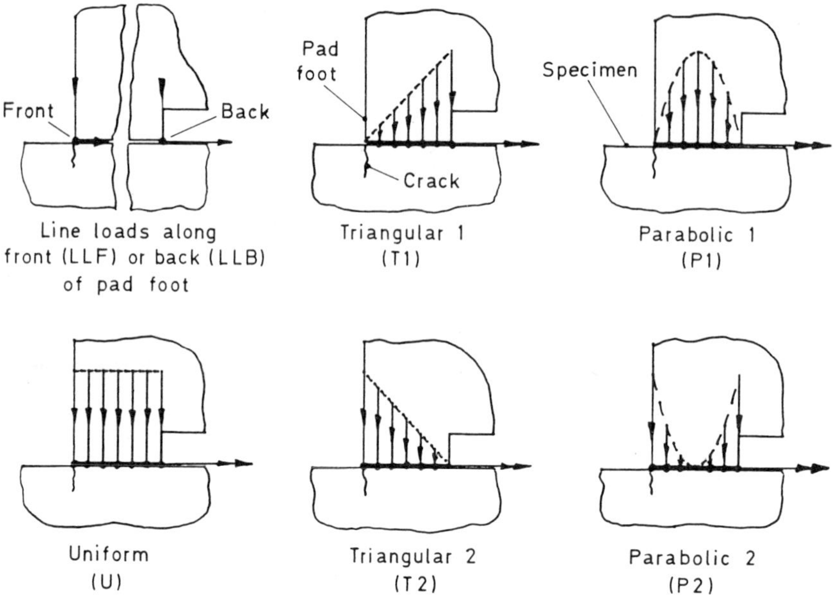

FIG. 3.2. Assumed distribution of frictional and normal pad loads.

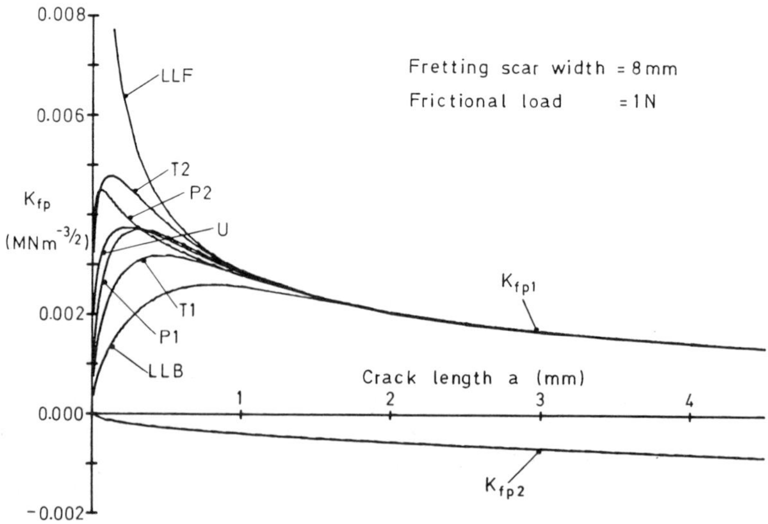

FIG. 3.3. Stress intensity factors for different distributions of frictional pad load.

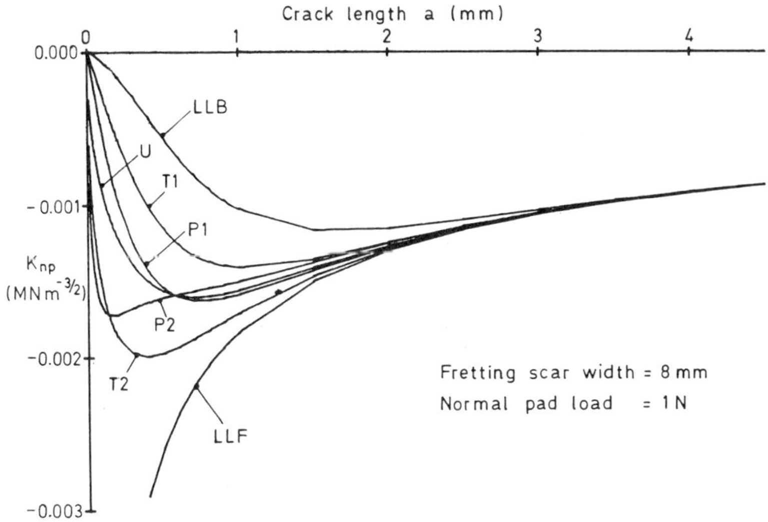

FIG. 3.4. Stress intensity factors for different distributions of normal pad load.

alternating body stress is lower under the pads because some load is diverted through the pads.

From reference 10:

$$K_{\mathrm{fp2}} = -1{\cdot}12\frac{F_{\mathrm{f}}}{A}\sqrt{\pi a} \tag{3.1}$$

The alternating stress intensity factor K_{a} is given by the components due to alternating body stress σ and a frictional force F_{f}.

$$K_{\mathrm{a}} = 1{\cdot}12\,\sigma_{\mathrm{a}}\sqrt{\pi a} + \frac{0{\cdot}008}{e}F_{\mathrm{f}}K_{\mathrm{fp1}} - 1{\cdot}12\frac{F_{\mathrm{f}}}{A}\sqrt{\pi a} \tag{3.2}$$

and the mean stress intensity factor by

$$K_{\mathrm{m}} = 1{\cdot}12\,\sigma_{\mathrm{m}}\sqrt{\pi a} + \frac{0{\cdot}008}{e}F_{\mathrm{n}}K_{\mathrm{np}} + K_{\mathrm{d}} \tag{3.3}$$

where K_{d} is the crack length correction stress intensity factor described in Section 3.4.3 and used in some predictions, and e is the fretting scar width.

The term $0{\cdot}008/e$ is included in the above equations to allow for the fact that scar widths vary somewhat with fretting conditions and K_{fp1} and K_{np} apply to a scar width of 8 mm.

Both mean and alternating stress intensity factors were multiplied by the finite width correction of Harris[16] for an edge crack in a specimen with the ends restrained in bending. This takes the form of a function of a/w:

$$f\left(\frac{a}{w}\right) = \frac{5}{1\cdot 12\left[20 - 13\left(\frac{a}{w}\right) - 7\left(\frac{a}{w}\right)^2\right]^{1/2}} \tag{3.4}$$

The above finite width correction was found to be the most appropriate one to use following some residual static strength tests on part-fatigued specimens.[14]

Figure 3.5 shows calculated peak stress intensity factors, not including the length correction K_d (Section 3.4.3), for a standard fretted specimen under constant amplitude loading at 35 MN m^{-2} rms alternating stress. One curve is shown for each of the assumed pad load distributions and also for a specimen without pads.

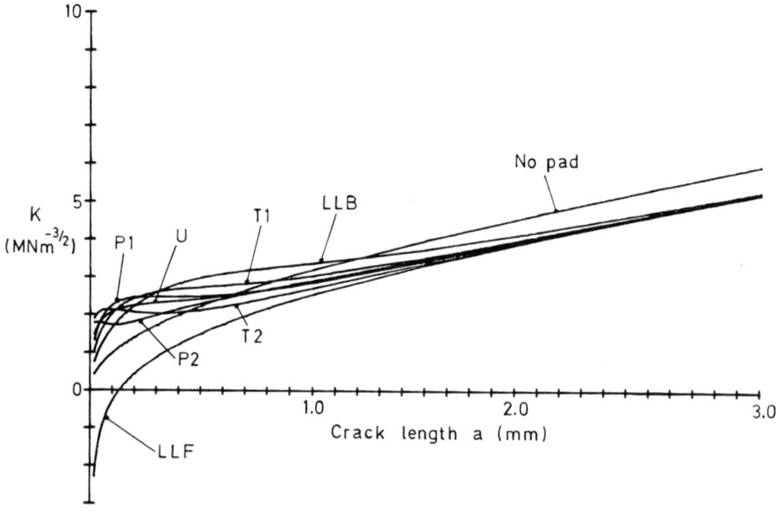

FIG. 3.5. Peak stress intensity factors for a range of pad load distributions.

As can be seen, all the fretting cases, except for assumed line loads at the front of the pad foot, show similar behaviour in that the stress intensity factor rises sharply just under the surface and then tends to level off. All the curves for fretting cases predict that for cracks longer than about 1 mm, stress intensities are lower than in the unfretted case.

Thus the fracture mechanics model is consistent with the observation of

Endo and Goto[7] that under fretting conditions, crack rate at short crack lengths is dependent on alternating frictional force. This leads to a general deduction that the damaging effect of fretting does not depend only on early crack initiation. Equally important are the raised stress intensities just below the surface which enable cracks to propagate from much smaller initial flaw sizes than for plain specimens, and speed up the crack at short crack lengths.

3.4.2 Calculation of Crack Rates

At specified values of crack length the mean stress intensity factor K_m was calculated from eqn (3.3), together with the appropriate range of alternating stress intensity factors K_a from eqn (3.2) corresponding to the alternating rms stress σ_r applied. Equation (3.2) requires values of measured frictional force, interpolated where necessary, as described in Section 3.5. Only one value of K_a per crack length was required for constant amplitude loading. This was used together with K_m and a computer subroutine[18] to obtain the crack rate $\mathrm{d}a/\mathrm{d}N$ from Pearson's[19] crack rate data for 'long' cracks (>0·25 mm). For variable amplitude loading at each crack length, crack rates were calculated for K_m and the range of values of K_a appropriate to the magnitudes of the individual cycles in the spectrum. The final rate was obtained by a linear summation of these rates weighted according to the probability of occurrence of K_a within each variable amplitude sequence, viz.

$$\frac{\mathrm{d}a}{\mathrm{d}N_{\mathrm{random}}} = \int_0^{\infty} p(K_a)\frac{\mathrm{d}a}{\mathrm{d}N}(K_a)\,\mathrm{d}K_a \tag{3.5}$$

Curves of crack length v. cycles were obtained by integrating numerically the inverse crack rate $\mathrm{d}N/\mathrm{d}a$ along the crack length. Failure was assumed to occur when the peak stress intensity factor exceeded K_{IC} which was taken to be 27·5 MN m$^{-3/2}$.[20] As can be seen from eqn (3.5) no allowance was made for retardation of the crack rate by high loads in the spectrum or for the difference in test frequency between that used for the crack propagation data (25 Hz) and that used in the test programme (120 Hz).

3.4.3 Use of the Crack Length Correction

A number of predictions in this chapter were made with the crack length correction (length correction) stress intensity factor[9,10] K_d included in the value of K_m. The origin of this is as follows.

Frost *et al.*[21] have shown that for a range of metals at zero mean stress the stress intensity factor giving the threshold of crack propagation at short crack lengths ($a < 0{\cdot}25$ mm) is only about 60% of that for crack lengths longer than 0·5 mm. Another way of expressing this is as the incompatibility between the observed empirical rule for determining the threshold of crack propagation for cracks over a range of lengths up to 4 mm

$$\sigma_c^3 a = \text{constant} \tag{3.6}$$

and the rule expected by the application of elementary fracture mechanics, i.e.

$$\sigma_c^2 a = \text{constant} \tag{3.7}$$

where σ_c is the alternating stress below which the crack will not propagate.

Also, Fisher and Sherratt[22] found, when monitoring crack growth from notches in mild steel specimens, that short cracks ($\approx 0{\cdot}1$ mm) propagated considerably faster than longer cracks ($\approx 0{\cdot}6$ mm) where the computed alternating stress intensity factor was the same in both cases. Similar results have been reported by Dowling.[23]

Pook[24] has suggested that the effect can be explained by crack closure such that at short crack lengths the plastic zone at the tip of the crack is more effective at propping open the crack than at longer crack lengths. Thus at short crack lengths the effective range of stress intensity factor is greater because the crack does not close until the crack is some way into compression. The length correction stress intensity factor K_d is an estimate of this extra effective range of stress intensity factor. It is given by the difference between the experimentally determined semi-range alternating stress intensity factor for the crack length under consideration and that for long cracks. It has been computed for L65 aluminium alloy[9,10] as

$$\begin{aligned} K_d &= 2{\cdot}09 - 5{\cdot}27a^{1/6}\ \text{MN m}^{-3/2} \qquad (a \leq 3{\cdot}84\,.\,10^{-3}\ \text{m}) \\ K_d &= 0 \qquad (a \geq 3{\cdot}84\,.\,10^{-3}\ \text{m}) \end{aligned} \tag{3.8}$$

It should be noted that a different form of length correction, also designed to account for anomalies at short crack lengths, has recently been proposed by Haddad *et al.*[25,26]

3.5 SUMMARY OF FRICTIONAL FORCE MEASUREMENTS

A series of measurements of frictional force between fretting pads and specimens was made under constant[11] and variable[12] amplitude loading. The measurements were made at the same test frequency as the rest of the tests described in this chapter, using specimens as illustrated in Fig. 3.1. There were, however, detail differences in the fretting pads, arising from the need to affix the strain gauges by means of which the frictional forces were measured. Fatigue lives achieved on specimens upon which frictional forces were measured were recorded and generally agreed reasonably well with those obtained in the rest of the test programme.

In the frictional force measurements, two types of slip (or relative movement between pads and specimens) were noted confirming previous observations by Endo and co-workers.[6] The first type was termed 'elastic slip'. Where this occurred, frictional forces were approximately proportional to slip, and fretting scars were relatively superficial. Following work by O'Connor and Johnson[27] it was concluded that in this situation frictional forces were developed by elastic deformation of contacting asperities. The second type of slip was termed 'macro-slip', during which the frictional force was relatively independent of slip. In this case the asperities were sliding over each other and the fretting pads tended to wear away the specimen surface thus forming a groove ('keying-in'[28]).

In all the tests carried out using constant amplitude loading it was found that there was a very short initial bedding-in period of elastic slip and macro-slip, during which alternating frictional forces increased rapidly. During this period macro-slip reduced until it either ceased altogether at less than 1 % of the fatigue life, or continued virtually throughout the life. In this latter situation keyed-in scars were noted and 'frictional' forces tended to increase sharply at the peaks and troughs of the loading waveform, indicating that the fretting pads were coming up against the sides of the keyed-in grooves. This suggested that where macro-slip was taking place, the frictional forces at the peaks of each loading cycle would tend to be concentrated at the back of each pad foot. Significant keying-in was found to occur wherever average semi-range coefficients of friction greater than 0·85 were recorded. Figure 3.6 summarises the frictional forces measured under constant amplitude loading. As can be seen, for the specimens with a pad load of 2·5 kN it was found that average frictional forces generally increased with increasing pad span due to the action of elastic slip. However, at the higher alternating stress levels on specimens with the two longest pad spans, substantial macro-slip occurred and frictional forces peaked at a coefficient of friction of approximately unity. On specimens with the low pad load, frictional forces also peaked at a coefficient of friction of approximately unity. It should be noted that this represented only 40 % of the maximum frictional forces recorded at the higher pad load.

Under variable amplitude loading, each loading sequence contained a

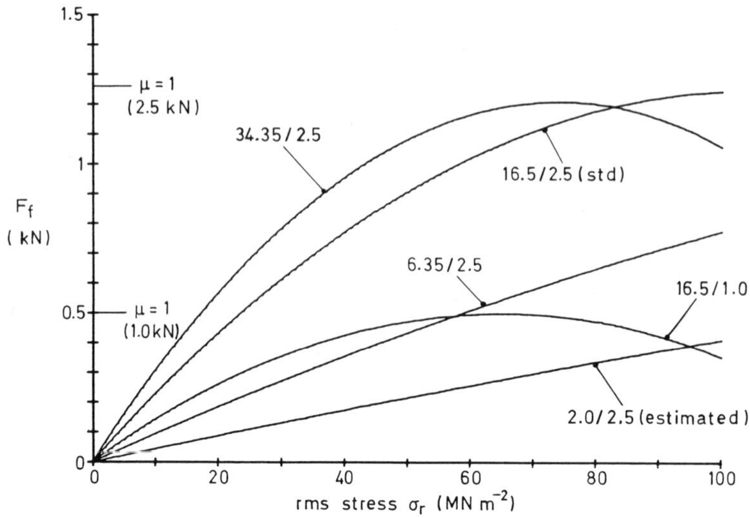

FIG. 3.6. Constant amplitude frictional forces at different pad spans (mm)/pad loads (kN).

range of component alternating stresses each with its associated range of alternating frictional force values and these were monitored simultaneously. Thus for any variable amplitude test a curve of the type shown in Fig. 3.6 could be constructed which applied to the component cycles. This curve was not in general the same as that obtained under constant amplitude loading for specimens with the same pad load and span, and in some cases different curves were obtained for different rms values of variable amplitude stress on specimens with the same pad load and span. This difference was associated with the bedding-in process which, under constant amplitude loading, rapidly increased the frictional force value at which macro-slip occurred and hence the maximum frictional force. Under variable amplitude loading, however, the bedding-in process was continually disturbed by groups of higher level cycles which reduced the frictional force at which macro-slip occurred. The bedding-in would then start again. The outcome of this was that keying-in was much more marked under variable amplitude loading. Also the steady state frictional forces measured under constant amplitude loading tended to be upper bounds for the values measured at component cycles in variable amplitude sequences. This is shown in Fig. 3.7 which compares frictional forces measured for standard fretted specimens under variable amplitude loading at three rms stress levels with those recorded in the constant amplitude tests. Because of

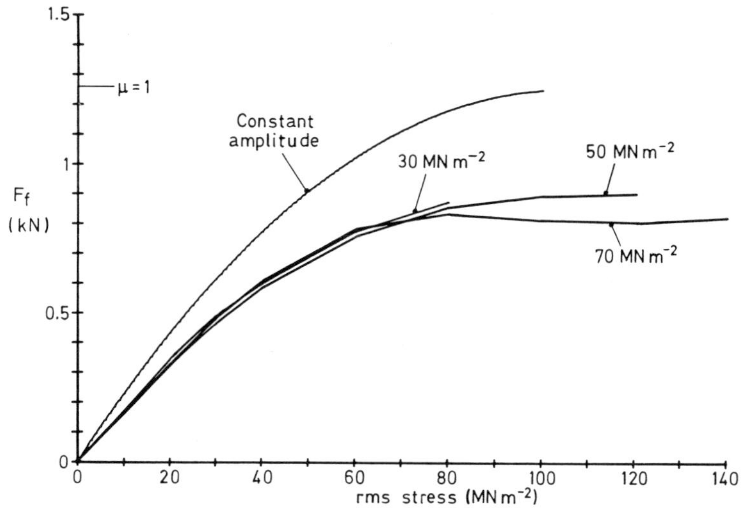

FIG. 3.7. Frictional forces on standard fretted specimens—constant and variable amplitude loading.

the continual changes in frictional force at which macro-slip occurred there was considerable scatter in measured frictional forces at the higher component rms levels. However, as can be seen, there was little variation with random rms level of the average curves.

Figure 3.8 compares frictional forces under the two types of loading for specimens with the same pad load as above but with the longer pad span of 34·35 mm. As can be seen on these specimens which showed the largest amount of macro-slip of any tested, there was a marked difference between the frictional forces at the three random rms levels. Also, frictional forces measured at the lower component rms alternating stress levels were similar to those under constant amplitude loading. Similar relative behaviour of frictional forces under the two loading actions was found for specimens with the low pad load.

Specimens with a pad span of 6·35 mm and a pad load of 2·5 kN showed frictional forces consistently about 70 % of those recorded under constant amplitude loading.

Thus it can be seen that there was no simple relationship between frictional forces and component alternating stresses within variable amplitude loading sequences. The relationship changed with the random rms alternating stress as well as with pad load and span. Therefore when applying eqn (3.2) in the fracture mechanics predictions, frictional forces

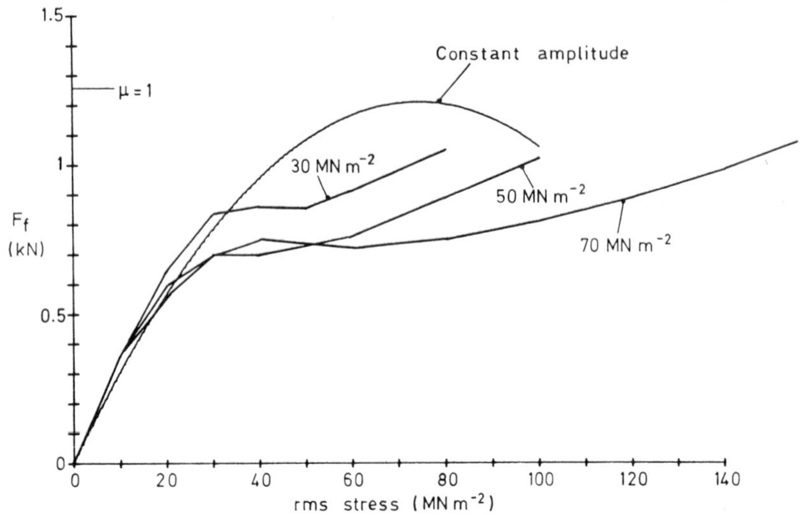

FIG. 3.8. Frictional forces for constant and variable amplitude loading—pad span = 34·35 mm; load = 2·5 kN.

were obtained by interpolating the appropriate measured values. Where frictional forces were required for random rms levels higher than 70 MN m^{-2} or lower than 30 MN m^{-2}, the values for 70 MN m^{-2} or 30 MN m^{-2} respectively were assumed.

3.6. FRACTURE MECHANICS PREDICTIONS

All the stress intensity factors used for the prediction described in this section were computed assuming fretting-scar widths as in Table 3.2. It was assumed in all cases that the distribution of normal loads had the same shape as the distribution of frictional loads.

3.6.1 Effect of Initial Flaw Size and Length Correction on Predicted Life

All fracture mechanics analyses have to assume an initial flaw size from which the crack propagates. Examination of the fracture surfaces of the fretted specimens showed that by the time cracks had reached a depth of 0·02 mm, the majority were propagating in stage II manner, although they tended to slant slightly so as to grow under the fretting pads.

Figure 3.9 compares experimentally achieved lives of standard fretted specimens under constant amplitude loading with those predicted assuming

TABLE 3.2
WIDTHS OF FRETTING SCARS

Pad load (kN)	*Span (mm)*	*Scar widths*	
		Constant amplitude (mm)	*Gaussian random (mm)*
2·5	34·35	8·0	9·5
2·5	16·50	8·0	8·1
2·5	6·35	8·0	8·5
2·5	2·00	8·0	8·1
1·0	16·50	5·6	7·0

uniform pad load distributions and initial flaw sizes of 0·02 and 0·1 mm with and without the length correction. Considering first the predictions without the length correction, these were less than satisfactory in two respects. First the predicted curves were too 'flat'. At the higher stress levels predicted, lives were too short and at the lower stress levels they were too long. Also the predicted fatigue limits were due to the initial flaw not propagating, whereas experimental evidence[1] was that the fatigue limit was due to cracks stopping

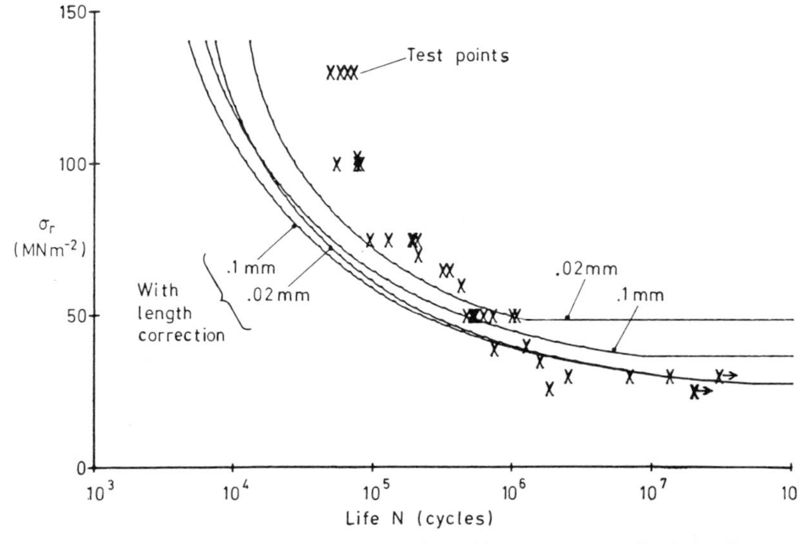

FIG. 3.9. Effect of initial flaw size on predicted life—constant amplitude loading.

at around 0·2 mm. Considering now the predictions using the length correction, it can be seen that the fatigue limit was predicted very much better and in fact the predicted fatigue limit was this time due to cracks stopping after an initial period of propagation. Also, the predicted curve was steeper than without the length correction, which was also an improvement. However, at the highest stress levels the predicted lives were too short by a large factor.

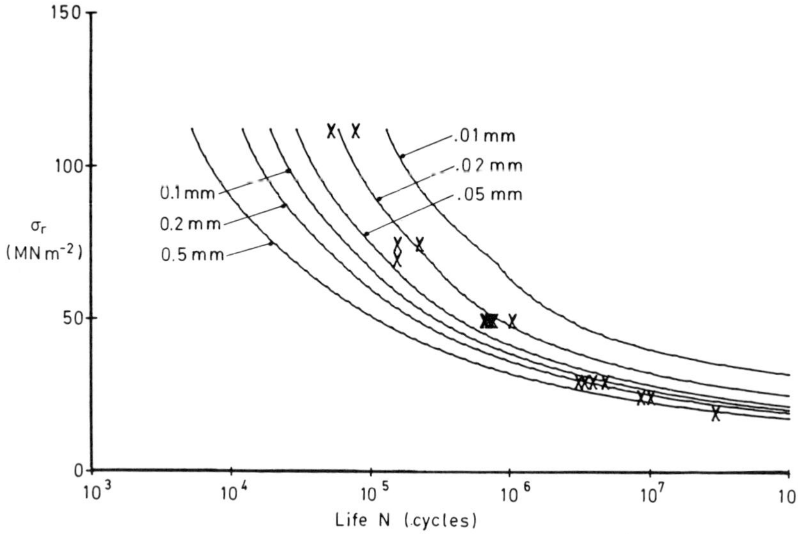

FIG. 3.10. Effect of initial flaw size on predicted life—variable amplitude; no length correction.

Figures 3.10 and 3.11 show the effect of initial flaw size on the predicted life of standard fretted specimens under variable amplitude loading without and with the length correction, respectively. Uniform pad load distributions were assumed. A comparison with Fig. 3.9 shows that the variable amplitude predictions were more accurate than the corresponding constant amplitude predictions. As before, the predictions without the length correction gave an *S–N* curve that was too flat and the use of the length correction gave predictions that were more accurate at the lower stress levels but predicted lives that were too short at the higher stress levels. However, the difference between predictions with and without the length correction was not as marked as for constant amplitude loading.

Figure 3.12 shows predicted variable amplitude *S–N* curves for plain

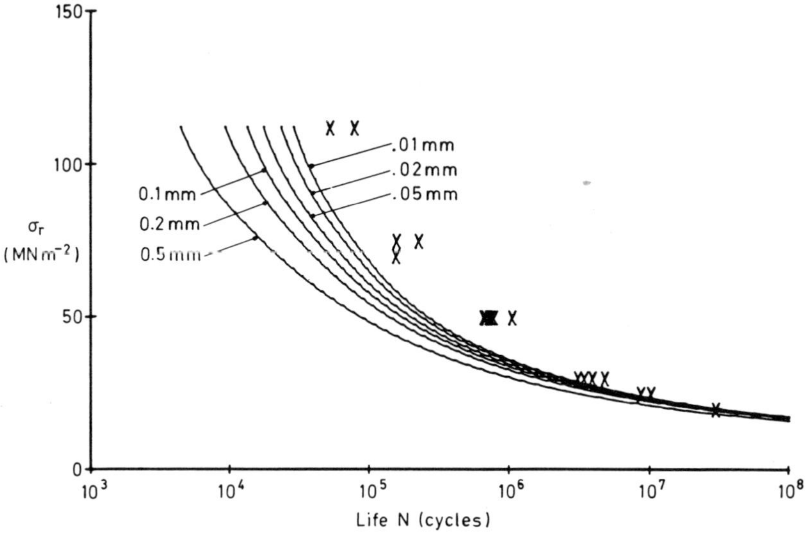

FIG. 3.11. Effect of initial flaw size on predicted life—variable amplitude; length correction used.

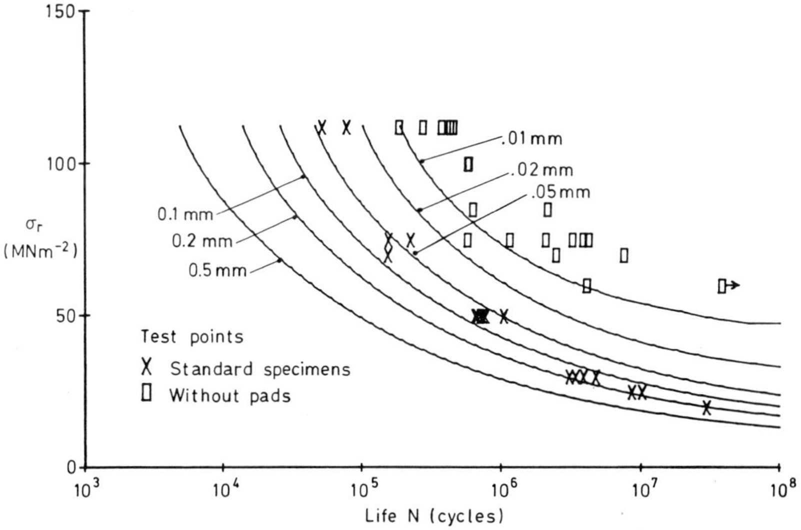

FIG. 3.12. Effect of initial flaw size on predicted life—no pads; variable amplitude; length correction used.

specimens assuming a range of initial flaw sizes and using the length correction. Experimental test points for standard fretted and plain specimens are shown for reference. Comparison with the equivalent predictions for standard fretted specimens (Fig. 3.11) shows that in the latter case the effect of initial flaw size was far less marked. The reason for this less critical behaviour for specimens with pads was that in this case the stress intensity factor, and hence the crack rate, was more nearly constant over the fatigue life (Fig. 3.5). The same was found to apply for constant amplitude loading.

Summarising the effect of initial flaw size on predicted fatigue life, it was found that it was not nearly as large for fretted as for unfretted specimens. The use of the length correction greatly improved predictions under both constant and variable amplitude loading close to the fatigue limit and also predicted the observed fact that the fatigue limit was due to cracks stopping after an initial period of propagation. However, the accuracy of prediction at the higher stress levels was worse with the length correction, particularly for constant amplitude loading.

3.6.2 Predicted and Measured Crack Propagation Curves

Figures 3.13 and 3.14 show predicted crack propagation curves for constant and variable amplitude loading on standard fretted specimens. In

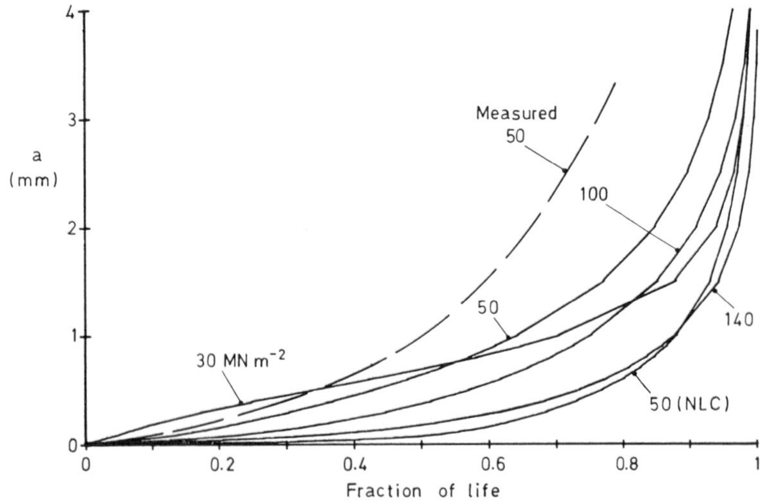

FIG. 3.13. Predicted and measured curves—constant amplitude loading.

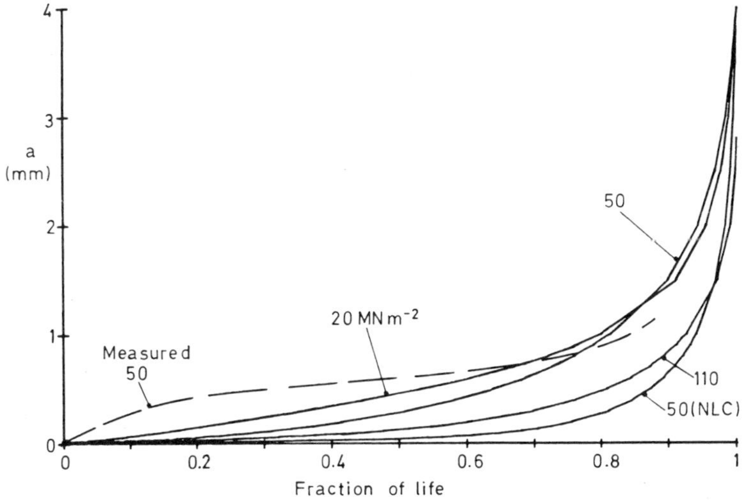

FIG. 3.14. Predicted and measured curves—variable amplitude loading.

both cases a uniform distribution of pad loads was assumed with an initial flaw size of 0·02 mm, and crack lengths were plotted against fraction of life consumed so that the shapes of the crack propagation curves could be compared at different stress levels. As can be seen in both these figures, the predictions, in which the length correction was used, varied markedly in shape with alternating stress level. Similar sets of curves for both loading actions, constructed without using the length correction, were found not to vary very much in shape with stress level. One curve predicted without the length correction is plotted for each loading action for an rms stress level of 50 MN m^{-2} and is labelled 'NLC'. Finally, a measured crack propagation curve for each loading action at an rms stress of 50 MN m^{-2} is plotted in Figs. 3.13 and 3.14. These curves were constructed by fatigue testing specimens to different percentages of the expected life to failure, breaking them statically and measuring the sizes of any fatigue cracks thus revealed.[14]

As can be seen from Fig. 3.13, the measured curve for constant amplitude loading agreed rather better with that predicted using the length correction than that without. Also, further similar measurements indicated that at 30 MN m^{-2} rms, cracks of 0·2 mm commonly existed at 5% of the life whereas at 62·5 MN m^{-2} the corresponding crack length was 0·06 mm. As can be seen from Fig. 3.13, although the predicted crack lengths at 5% of the life were not as long as the measured values above, the predicted curves using the length correction were more nearly in accordance with the measurements in that at the lower stress levels the crack reached any particular length at a much earlier percentage of the life than at the higher stress levels.

For the case of variable amplitude loading it can be seen that although the prediction using the length correction gave better agreement with the measured curve than that without, the agreement was not good. In particular, neither of the predictions indicated the observed fact that the crack curve would start off at a relatively fast rate and then slow down. The overall shape of the measured curve was confirmed in other work. Standard fretted specimens were subjected to tests in which the fretting pads were removed at different stages in the life[1] and the tests continued to failure. It was found that under constant amplitude loading over a range of stress levels, removing the pads at up to 30 % of the life gave an increase in the life. However, for variable amplitude loading the corresponding value was 5 %. This difference in behaviour is explicable if it is accepted that under variable amplitude loading, but not under constant amplitude loading, fatigue cracks propagate relatively fast out of the region in which pad loads give a big contribution to the stress intensity factor and then slow down. A second piece of evidence confirming the shapes of the measured crack propagation curves involved acoustic emission traces obtained on standard fretted specimens under both types of loading.[14] Thirdly, under variable amplitude loading it was found that there was an initial period of acoustic emission activity which could have been associated with an early period of relatively fast crack growth. This was not found for constant amplitude loading.

These three separate pieces of evidence confirm that under variable amplitude loading the shape of the crack propagation curve was markedly different from that under constant amplitude loading. It would appear that in these specimens a fracture mechanics model which was reasonably accurate in predicting the shape of the crack propagation curve under constant amplitude loading was very inaccurate under variable amplitude loading. Therefore, either the stress intensity factors calculated for variable amplitude loading were much less accurate than for constant amplitude loading, or the cumulative damage model used was inaccurate. As described in Section 3.4.2, the cumulative damage model used was a linear summation of constant amplitude crack rates weighted according to the probabilities of occurrence of alternating stress intensity factors within each variable amplitude sequence.

With regard to the first possibility for error, as described in Section 3.4, account was taken of the main parameter which governs the magnitude of the stress intensity factor under fretting conditions, the frictional forces. It was thought that these could well be different under variable amplitude loading as in fact was the case (Section 3.5). Stress intensity factors were calculated using values measured under variable amplitude loading. However, although the resulting distributions of stress intensity factor gave predicted lives significantly different from those assuming the same frictional forces as under constant amplitude loading, the shapes of the distributions of stress intensity factor were similar to those used before. Also, the shapes of crack propagation curves predicted in Figs. 3.13 and 3.14 were similar. The only way that frictional forces could have explained the observed crack propagation curve for variable amplitude loading would have been if they had reduced drastically over the early part of the life. This was not found to be the case. An assessment of other parameters known to affect the magnitude of the stress intensity factor, e.g. pad load distribution, did not reveal any which would be capable of changing sufficiently under variable amplitude loading to account for the effect.

The other possible explanation for the unusual shape of crack propagation curves is that the cumulative damage model was not adequate. It is known that under variable amplitude loading the linear summation model used in this chapter normally predicts crack rates which are too fast. This has been attributed[29] to premature crack closure following the application of high loads in sequence. However, very little is known about crack propagation at short crack lengths. As discussed in Section 3.4.3, faster than expected crack rates at short crack lengths under constant amplitude loading have been attributed to a large difference between crack closure behaviour at short crack lengths and closure at long crack lengths. It is possible therefore that the cumulative damage behaviour in crack propagation may be different at short crack lengths in the region where a relatively accelerated growth rate was measured under variable amplitude loading.

Thus although the shape of the crack propagation curve was predicted better using the length correction than not using it, there were still substantial errors in the prediction. More work involving the measurement of cracks under constant and variable amplitude loading is necessary to establish the scope and magnitude of these errors and their cause. If these errors are shown to be widespread and as large as indicated by the limited data obtained at 50 $MN\,m^{-2}$, then the fracture mechanics model will need some modification.

3.6.3 Effect of Assumed Distributions of Pad Load on Fatigue Life

As was discussed in Sections 3.6.1 and 3.6.2, the predictions using the length correction described better the crack propagation behaviour of the fretted specimens, particularly for constant amplitude loading. Because of this, the remainder of the predictions presented in this chapter were made using the length correction. A more extensive assessment of the length correction is given in references 10 and 13.

As described in Section 3.6.1, the most inaccurate predictions made for standard fretted specimens assuming a uniform distribution of pad loads were under constant amplitude loading at the highest stress levels. However, there was some evidence (Section 3.5) that frictional forces at alternating stress peaks under these conditions were concentrated towards the back of the fretting pad foot, since it appeared that they arose from pad feet running into lines of debris or the edges of keyed-in grooves. This only happened at stresses above 70 MN m^{-2}, where keying-in was taking place. Figure 3.15 shows predicted and achieved lives for standard fretted specimens under constant amplitude loading with an assumed initial flaw size of 0·02 mm and a range of assumed pad load distributions. As can be seen, all the distributions which were symmetrical about the middle of the fretting pad foot predicted lives that were similar to those for uniform pad load distributions. It can also be seen that the assumption of distributions

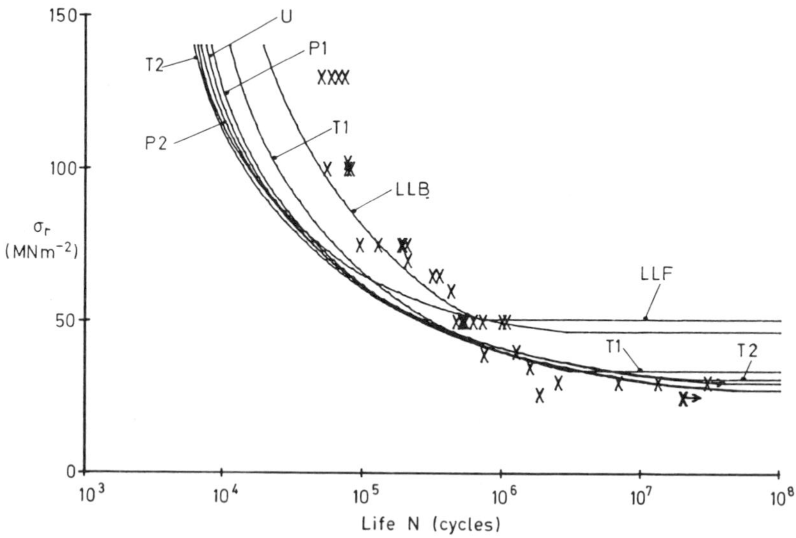

FIG. 3.15. Predicted lives for a range of pad load distributions—constant amplitude.

where the pad loads were more concentrated towards the back of the pad foot improved predictions at the higher stress levels. However, even with all the loads concentrated at the back of the pad foot, the predicted lives at the highest levels were still too short.

The predicted curve assuming line loads at the front of the pad foot can be seen from Fig. 3.15 to have a shape rather different from the others. Although not obvious from the predictions shown in this chapter the above assumption predicted some extreme changes in life as frictional and normal forces were varied. This was due to the fact that for this pad load distribution both the frictional and normal contributions to the stress intensity factor tended to infinity as zero crack length was approached (Figs. 3.3 and 3.4). Thus the stress intensity factor at short crack lengths was primarily the result of two large components that tended to oppose each other and a small percentage change in one component could give a large change in the resultant stress intensity factor. This extreme behaviour was not found for the other assumed pad load distributions which, as discussed above, were more likely to apply in practice.

Figure 3.16 shows the same set of predictions as for Fig. 3.15, except that Fig. 3.16 applies to variable amplitude loading. It can be seen that apart from the curve assuming a line load at the front of the pad foot (see above), the pattern of predictions was similar to that under constant amplitude

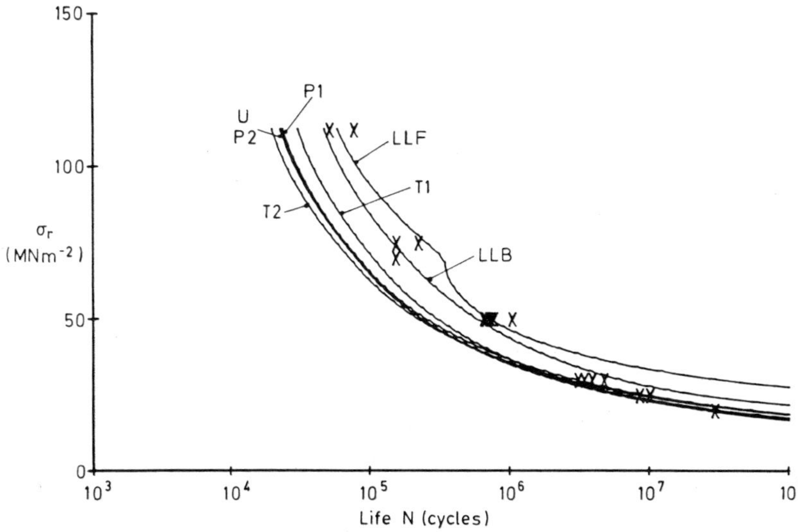

FIG. 3.16. Predicted lives for a range of pad load distributions—variable amplitude.

loading. However, particularly at the higher stress levels, the predictions tended to be more accurate. Under variable amplitude loading, keying-in occurred at all stress levels on the standard fretted specimens. Under these conditions the sudden build up in 'frictional forces' at the waveform peaks and troughs was evident, as for constant amplitude loading (Section 3.5), but was not as marked and only occurred at the higher level cycles within the sequence, presumably because only these cycles deflected the specimen enough for the pad feet to approach the sides of the keyed-in groove. Thus some tendency would be expected for the pad loads at the peaks of the stress cycles to concentrate towards the back of the pad foot, but not to the same extent as under constant amplitude loading. The most appropriate predicted curve would be expected to be between triangular 1 and the left hand group of curves representing symmetrical distributions. This would still give variable amplitude predictions that were rather more accurate than those for constant amplitude loading. However, two more factors would be expected to improve the variable amplitude predictions further.

First, in the variable amplitude case the frictional forces assumed for rms stresses above 70 MN m^{-2} were those measured at that level (Fig. 3.7). Had the fatigue machine on which the measurements were made been able to apply stresses above this level, it is likely that lower frictional forces would have been recorded (cf. Fig. 3.8) and predicted lives would have been longer. Secondly, allowance for retardation of the crack due to high loads in the sequence[30] would have further increased predicted life.

Thus variable amplitude predictions, particularly those taking into account the additional factors described above, were more accurate than the corresponding constant amplitude predictions. This was a rather surprising result in view of the errors indicated by a comparison of predicted and measured crack curves in Section 3.6.2.

3.6.4 Predictions for a Range of Pad Spans

Figures 3.17 and 3.18 show predictions for the two loading actions applied to specimens having the full range of pad spans at a pad load of 2·5 kN. The predictions were all made on the same basis, i.e. using the length correction, an initial flaw size of 0·02 mm and triangular 1 pad load distributions. As discussed in Section 3.6.3, there was some evidence that distributions of pad load varied depending on the fretting conditions and taking account of this would be expected to improve the predictions. However, the assumption of triangular 1 pad load distribution can be regarded as a reasonable compromise to illustrate the way that the model predicts the effect of pad span on fatigue life.

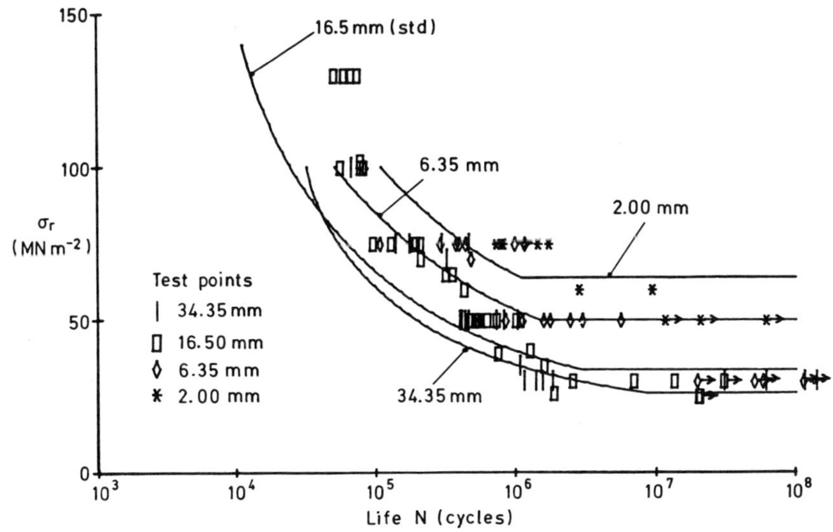

FIG. 3.17. Predicted and achieved lives for a range of pad spans—constant amplitude.

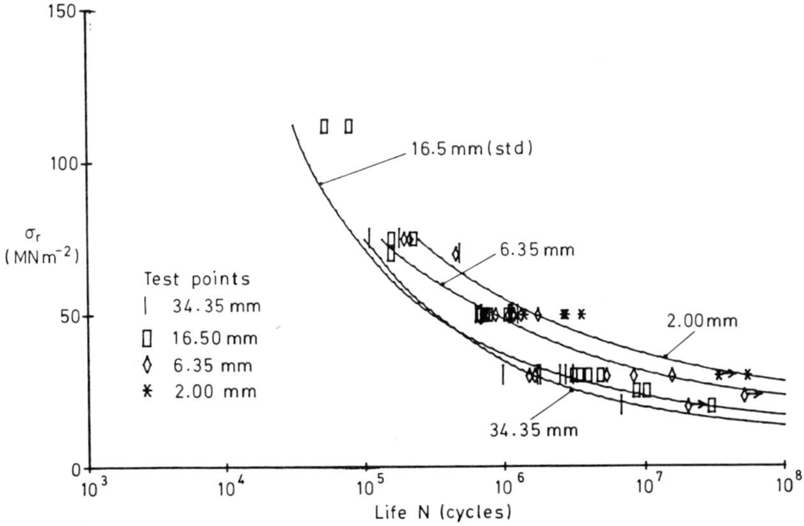

FIG. 3.18. Predicted and achieved lives for a range of pad spans—variable amplitude.

As can be seen from Figs. 3.17 and 3.18, the relative variation of fatigue life with pad span was predicted very well for constant and variable amplitude loading respectively. This predicted variation in life with pad span was due entirely to the associated changes in frictional force and the related stress intensity factor. Predicted lives were consistently more accurate under variable amplitude loading.

3.6.5 Predictions at Two Pad Loads

Figure 3.19 shows experimental and predicted lives for specimens with pad span of 16·5 mm and two values of pad load, 1 kN and 2·5 kN, under constant amplitude loading. Figures 3.20 shows the same for variable amplitude loading. Predictions were made making the same assumptions regarding initial flaw size etc. as described in Section 3.6.4. As can be seen, the relative effect of changing the pad load was predicted well for both constant and variable amplitude loading. As before, the most accurate life predictions were for variable amplitude loading. The effect of pad load on predicted and achieved fatigue life was small at the lower stress levels as can be seen for both constant (Fig. 3.19) and variable (Fig. 3.20) amplitude loading. This may appear to be rather surprising at first sight in view of the relatively large differences in frictional force recorded between the two pad

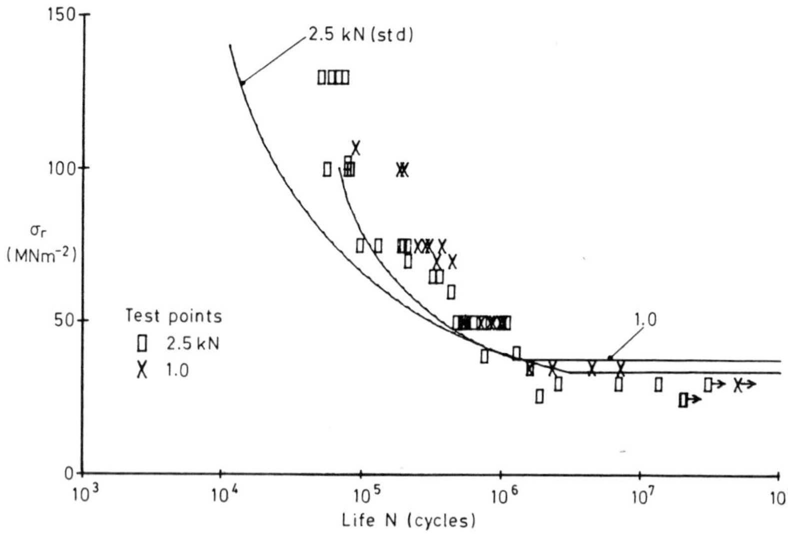

FIG. 3.19. Predicted and achieved lives for two pad loads—pad span = 16·5 mm; constant amplitude.

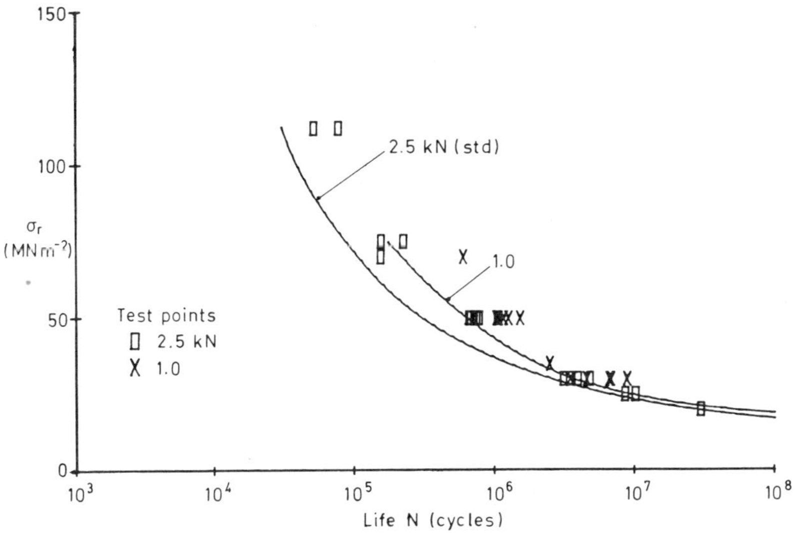

FIG. 3.20. Predicted and achieved lives for two pad loads—pad span = 16·5 mm; variable amplitude.

loads (Section 3.5). However, there were two factors which offset this. First the fretting scar width was less at the low pad load (Table 3.2). Much more important was the fact that reducing the normal pad load reduced the magnitude of its associated beneficial stress intensity factor. This tended to offset the reduction in the damaging stress intensity factor associated with the frictional loads. The net effect was that predicted lives could go up or down with pad load. As can be seen from Fig. 3.19, at different stress levels it was predicted that the life would generally increase with decrease in pad load, except over a small range of alternating rms stresses around 40 MN m^{-2} where a small decrease was predicted.

3.7 CONCLUDING DISCUSSION

It was concluded that the accuracy of the predictions under both constant and variable loading was very good considering the possible sources of error. There were some large errors in predicted lives at the higher stress levels under constant amplitude loading but these may well be reduced considerably as the model is refined to account for the sources of error listed below. Particularly encouraging was the way in which the model predicted the effect on life of varying the pad span and pad load for both constant and variable amplitude loading, and predicted shapes of crack propagation curves under constant amplitude loading which agreed well with limited available measured data. It is considered that the most serious deficiency in the model was its failure to predict the observed fact that under variable amplitude loading the crack propagated initially at a relatively fast rate and then slowed down. This is discussed in detail in Section 3.6.2. Possible sources of error which apply to predictions under both constant and variable amplitude loading are listed below:

1. Crack rate data for crack rates below about 3×10^{-5} mm/cycle is very scarce and numerical data defining the behaviour of cracks at lengths well below 1 mm is even scarcer. Pearson's data,[19] which was used in this chapter and assessed as the best available, covered a range of crack rates down to about 3×10^{-6} mm/cycle and was thereafter extrapolated over more than two decades of crack propagation rate down to the threshold value measured by Frost *et al.*[21] about 19 years ago. As discussed in Section 3.4.2, crack propagation in the region of this extrapolation is not well understood. Therefore since large proportions of the predicted lives in this chapter were calculated at short crack lengths with stress intensities close to the threshold of crack propagation, substantial errors were possible. The length correction was used in this chapter in an attempt to account for anomalous behaviour of short cracks and on the whole improved predictions, but it should be emphasised that it was based on limited data.
2. It is to be expected that local variations in the normal and frictional loads under the pad would lead to local variations of stress intensity factor close to the surface and locally affect crack propagation rates at short crack lengths.
3. Multiple cracks were present on many of the fretted specimens, particularly those in the keying-in situation and would tend to slow crack rates where one crack was shielding another.
4. Mode II stress intensity factors were present due to both normal and frictional pad loads and were not taken into account, due to lack of appropriate crack propagation data; neither was the fact that cracks tended to propagate initially at an angle so as to grow under the pads.
5. No allowance was made for the difference in frequency between that at which the crack rate data was obtained (25 Hz) and that at

which the fretting tests were carried out (120 Hz). However, available data on similar material[31] suggests that if frequency were accounted for, predicted lives could be increased by a factor of between 1·2 and 2.

6. It was assumed in all cases that the distributions of normal and frictional pad loads were the same. This may not have been the case, particularly when pads were keying-in. Also, pad load distributions could change throughout the life, again particularly when keying-in was occurring. Nevertheless, it should be remembered that the effect on life of pad load distribution was surprisingly small.

3.8 FURTHER WORK

3.8.1 Frictional force measurements and associated fatigue tests are continuing under a range of conditions including ranges of frequency and mean stress. The case of changes in mean stress is an interesting one because different fretted components and materials can show widely varying changes in life with change in mean stress.[32] The fracture mechanics model is potentially capable of explaining many of these as follows. When the mean body stress in a component changes, the associated stress intensity factor also shows an increase in mean value. However, the stress intensity factors associated with the fretting forces do not change with mean stress. Therefore, the model predicts that the early part of the life which is dominated by the fretting force components is little affected by the mean stress. Thus components with different combinations of fretting forces and body stress show different sensitivity to mean stresses.

3.8.2 It is intended to carry out crack propagation measurements at short crack lengths in specimens without fretting. The specimens will be subjected to constant and variable amplitude loading. It is hoped that an improved definition of cumulative damage behaviour in crack propagation at short crack lengths will lead to a better understanding of fretting.

3.9 CONCLUSIONS

3.9.1 A fracture mechanics analysis of aluminium alloy specimens with steel fretting pads, subjected to constant and variable amplitude loading at zero mean stress, has been carried out. Using measured data on frictional forces between the fretting pads and the specimens, predictions of fatigue lives were made for specimens having four different values of pad span (different values of slip) and two values of pad load.

3.9.2 The accuracy of predicted lives was good, particularly for variable amplitude loading. The most inaccurate predictions were made for constant amplitude loading at the highest alternating stresses. The effect on fatigue life of changing pad span and load was predicted well.

3.9.3 A crack length correction was assessed which empirically took account of the fact that short cracks having the same alternating stress intensity factor as long cracks tend to propagate faster. It was found that the use of the length correction gave very good predictions of the fatigue limits, and, overall, gave the best predicted *S–N* curve shapes. It also gave predicted crack propagation curves which best agreed with measured data.

3.9.4 The most serious failure of the fracture mechanics model was that it did not predict the observed fact that under variable amplitude loading there was an initial relatively fast rate of crack propagation after which it slowed down.

REFERENCES

1. Edwards, P. R. and Ryman, R. J. Studies in fretting fatigue under service loading conditions. *Proc. Eighth ICAF Symposium*, Lausanne (1975); also RAE TR 75123 (1975).
2. Edwards, P. R. Cumulative damage in fatigue with particular reference to the effects of residual stresses. RAE TR 69237 (1969); (ARC CP 1185) (1969).
3. Milestone, W. D. and Janeczko, J. D. Friction between steel surfaces during fretting. *Wear*, **18**, 29–40 (1971).
4. Wharton, M. H., Waterhouse, R. B., Hirakawa, K. and Nishioka, K. The effect of different contact materials on the fretting fatigue strength of an aluminium alloy. *Wear*, **26**, 253–60 (1973).
5. Endo, K., Goto, H. and Nakamura, T. Fretting fatigue strength of several materials combinations. *Bull. JSME*, **17**(92) (1973).
6. Endo, K., Goto, H. and Fukunaga, T. Behaviours of frictional force in fretting fatigue. *Bull. JSME*, **17**(108) (1974).
7. Endo, K. and Goto, H. Initiation and propagation of fretting fatigue cracks. *Wear*, **38**, 311–24 (1975).
8. Fenner, A. J. and Field, J. E. Fatigue under fretting conditions. NEL AB Div 16/57 (1957).
9. Edwards, P. R., Ryman, R. J. and Cook, R. Fracture mechanics prediction of fretting fatigue. *Proc. Ninth ICAF Symposium*, Darmstadt (1977).
10. Edwards, P. R., Ryman, R. J. and Cook, R. Fracture mechanics prediction of fretting fatigue under constant amplitude loading. RAE TR 77056 (1977).
11. Edwards, P. R. and Cook, R. Frictional force measurements on fretted specimens under constant amplitude loading. RAE TR 78019 (1978).

12. Edwards, P. R. and Cook, R. Frictional force measurements on fretted specimens under variable amplitude loading. RAE TR 78059 (1978).
13. Edwards, P. R. and Cook, R. Fracture mechanics prediction of fretting fatigue under Gaussian random loading. RAE TR 78086 (1978).
14. Ryman, R. J. and Blackwell, M. P. An acoustic emission investigation of the initiation and propagation of fretting fatigue cracks in BS L65 aluminium alloy under constant and random amplitude loading. Testwell Ltd. Report FAT/130, Issue 2 (1978).
15. Edwards, P. R. A controller for random and constant amplitude loading on resonant and electrohydraulic fatigue machines. RAE TR 66023 (1966).
16. Rooke, D. P. and Cartwright, D. J. Compendium of stress intensity factors. HMSO, London (1974).
17. Rooke, D. P. and Jones, D. A. Stress intensity factors in fretting fatigue. RAE TR 77181 (1977).
18. Edwards, P. R. A computer program for the interpolation and extrapolation of crack propagation data. RAE TR 76115 (1976); (ARC CP 1387, 1977).
19. Pearson, S. Initiation of fatigue cracks in commercial aluminium alloys and the subsequent propagation of very small cracks. RAE TR 72236 (1972).
20. Pearson, S. Fatigue crack initiation and propagation in half-inch (12·7 mm) thick specimens of aluminium alloy. RAE TR 71109 (1971).
21. Frost, N. E., Pook, L. P. and Denton, K. A fracture mechanics analysis of fatigue crack growth data for various materials. NEL/A2/1/69 (1969).
22. Fisher, B. C. and Sherratt, F. A fracture mechanics analysis of fatigue crack growth data for short cracks, in *Fracture mechanics in engineering practice*. P. Stanley, ed. Applied Science Publishers, London (1977).
23. Dowling, N. E. Crack growth during low cyclic fatigue of axial specimens. Westinghouse labs paper PALFA-P2 (1976); also, ASTM STP 637 (1978).
24. Pook, L. P. Various aspects of the fatigue damage threshold in mild steel. *Proc. SEE Conference 'Fatigue Testing and Design'*, London (1976).
25. Haddad, M. H. E. L., Smith, K. N. and Topper, T. H. Fatigue crack propagation of short cracks. ASME Paper 78-Mat-7 (1978).
26. Haddad, M. H. E. L., Topper, T. H. and Smith, K. N. Prediction of non-propagating cracks. *Engineering Fracture Mechanics*, **11**, 573–84 (1979).
27. O'Connor, J. J. and Johnson, K. L. The role of surface asperities in transmitting tangential forces between metals. *Wear*, **6**, 118–39 (1963).
28. Field, J. E. and Walters, D. M. Fretting fatigue strength of EN26 steel. Effects of mean stress, slip and clamping conditions. NEL Report 275 (1967).
29. Elber, W. Fatigue crack closure under cyclic tension. *Engineering Fracture Mechanics*, **2**, 37–45 (1970).
30. Sippel, K. O. and Weisgerber, D. Flight by flight crack propagation test results with several load aspects and comparison with calculation according to different models. *Proc. Ninth ICAF Symposium*, Darmstadt (1977).
31. Hartman, A. and Schijve, J. The effects of environment and load frequency on the crack propagation law for macro fatigue crack growth in aluminium alloys. *Engineering Fracture Mechanics*, **1**, 615–31 (1970).
32. Perrett, B. H. E. Fatigue endurance of structural elements in various materials under constant and variable amplitude loadings. RAE TR 77162 (1977).

4

Occurrence of Fretting Fatigue Failures in Practice

P. J. E. Forsyth

Materials Department, Royal Aircraft Establishment, Farnborough, Hants, UK

4.1 INTRODUCTION

In this chapter it is proposed to discuss the circumstances under which fretting fatigue failures arise in service, and to describe the diagnostic features that can be observed.

Fretting fatigue occurs when repeated loading on a structure or part causes sliding movement at metallic interfaces in the design. This movement, in turn, must induce, at or near certain points of contact between the faces, fatigue stresses of sufficient intensity to cause cracking. Thus there are certain key requirements for failure:

1. some degree of freedom for interface movement;
2. individual points of contact at asperities;
3. the development of, at or near these points, a sufficiently high fatigue stress to cause cracking.

There are numerous engineering situations where these requirements may be encountered. The more common ones are generally known to designers, but others arise which have not been envisaged at the design stage and yet can be equally disastrous. It is, perhaps, this latter class that demands particular attention, because in revealing such a diversity of possibilities, these failures demonstrate how alert designers must be to the dangers.

It is proposed to deal in the first instance with the identification of true fretting fatigue, and then describe the more commonly observed cases. This will be followed by consideration of more specific engineering situations where failures still commonly occur, but where the fretting aspect is at least recognised. The latter part of the chapter will be devoted to several case histories where somewhat unusual and unexpected failures have been experienced, and where the question can be fairly asked, should these events have been foreseen at the design stage?

4.2 THE IDENTIFICATION OF TRUE FRETTING FATIGUE

Before describing individual cases it is necessary to identify, as far as is possible, the features of a fretting initiated fatigue failure that unequivocally prove it to be such.

The rubbing together of clean metallic surfaces soon produces what are frequently described as scars at some of the surface asperities. These scars characteristically reveal the direction in which the rubbing action has been

FIG. 4.1. Elongated fretting scars indicating direction of relative motion BS L65 aluminium alloy. ×3·5.

operating by being streaked and elongated in that direction, as shown in Fig. 4.1. The scar generally has a roughened appearance, although in some instances it may be so compacted as to form a glaze. Scars on steels have a characteristic red–brown colour and on aluminium alloys they are black; they may also vary from a compacted glaze to a powdery deposit. In cases of excessive fretting of steel parts the red powdery dust, often described as 'cocoa', may be blown and deposited away from the site of its production. Fretting products are oxides of the metals concerned often with finely divided parent metal present, and in some cases pick-up or transfer occurs between the mating metal surfaces. The roughening of the metal surface itself, that is associated with the fretting product, may vary from patches of microscopic pits to large, shallow abraded areas. Fretting product may become compacted in the mouth of the crack, although such a feature is not in itself evidence of fretting-induced cracking, and this raises the important point of cause and effect, bearing in mind that the loosening of a joint by fatigue can itself cause considerable fretting. This may mask the actual evidence of pre-crack fretting, i.e. true fretting fatigue.

These various features and aspects of the problem will be dealt with in greater detail in relation to the cases described.

4.3 COMMON ENGINEERING SITUATIONS WHICH MAY SUFFER FROM FRETTING FATIGUE

There are two clearly distinguishable classes of fretting fatigue related to fastened joints.

1. Flanges, i.e. flat faces in contact subjected to varying shear loads, which, depending on the normal closure loads on the faces, may suffer some degree of slippage.
2. Loaded, filled holes, i.e. pins or fasteners bearing on the bores of holes and subjected to some interface sliding and to varying interface pressures.

Most cases of fretting can be resolved into versions of one or the other of these two classes.

The reason for making a distinction between classes 1 and 2 is not that the local physical process of fretting will be different, but that in the flange case the tangential and normal loads operate virtually independently of one another, whereas in the second case they are interdependent.

4.4 OTHER DESIGN DETAILS THAT CAN GIVE FRETTING FATIGUE PROBLEMS

1. Contacting strands in wire ropes or cables.
2. Spring washers of various forms, and circlips.
3. Leaf springs and coil springs.
4. Keys in keyways on shafts, and splines.
5. Wheel bosses on shafts, and collars and bushes.

4.5 SPECIFIC EXAMPLES OF FLANGE FACE FRETTING FATIGUE

It was mentioned earlicr that fretting scars appear at asperities at mating surfaces where intimate metal-to-metal contact is made. It follows from this that any lack of flatness of the component parts that may be introduced during manufacture or assembly is likely to influence the position of these scars. This is more likely to be so with thin-sheet construction where inadvertent distortion may be introduced in the cutting and forming operations. Fastener closure itself can cause appreciable dishing around the holes that results in non-uniform contact. These various situations are illustrated in Fig. 4.2. It is also common to find asperities in the form of burrs around holes that result from drilling, and in these situations the paint barrier may be broken and metal-to-metal contact points are formed as illustrated in Fig. 4.3.

With unpainted parts, such features as described would be very likely to

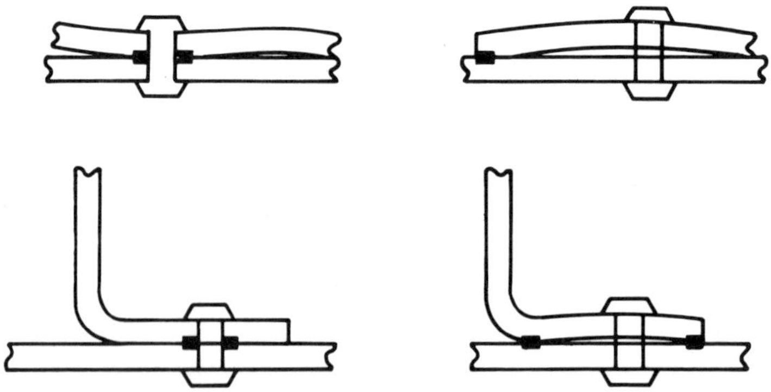

FIG. 4.2. Schematic diagrams of thin sheet joints illustrating commonly occurring fretting points. Danger points indicated thus ■.

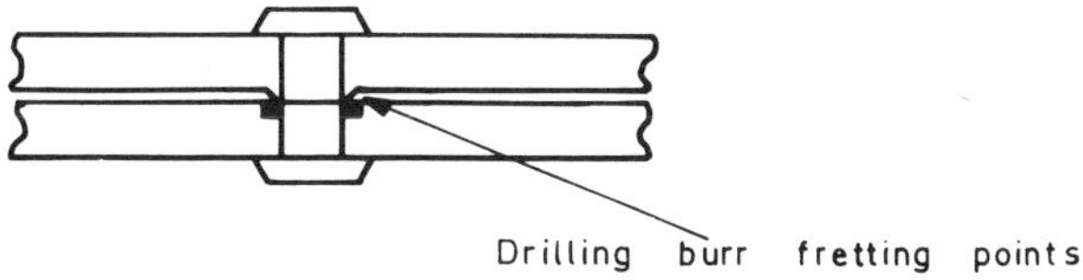

FIG. 4.3. Schematic diagram illustrating how drilling burrs cause fretting points. Danger points indicated thus ■.

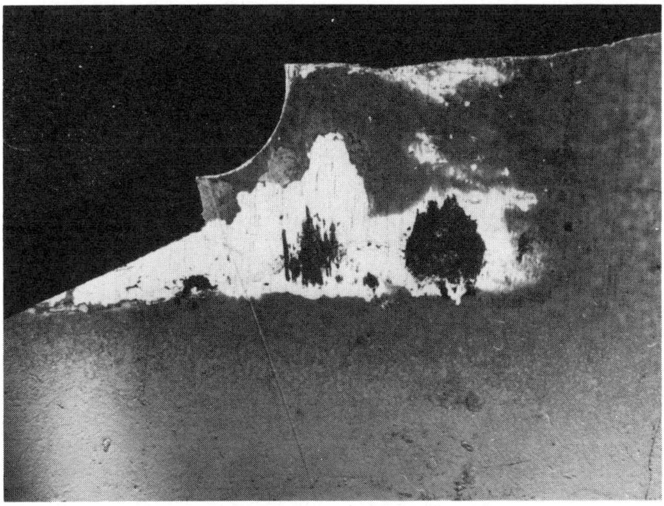

FIG. 4.4. An anodised and painted aluminium alloy surface where fretting abrasion has penetrated to cause metal-to-metal contact. ×4.

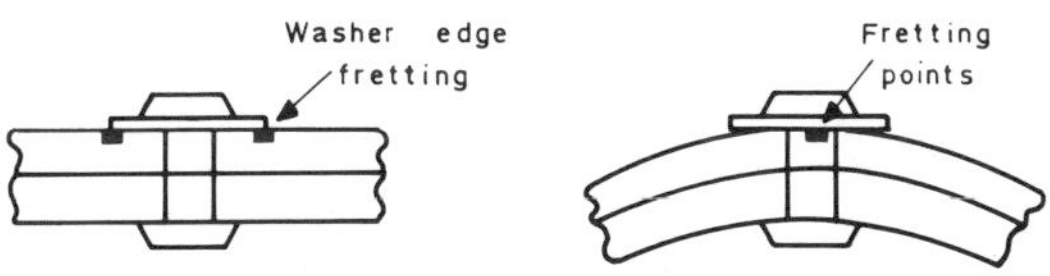

FIG. 4.5. Schematic diagram illustrating the positions of frequently observed fretting damage. Danger points indicated thus ■.

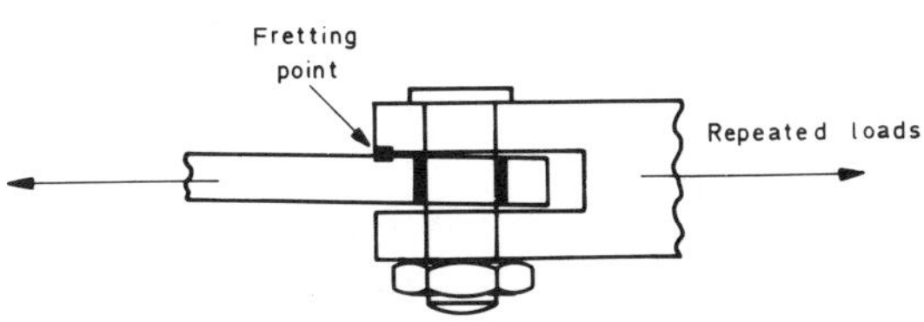

FIG. 4.6. Schematic diagram illustrating how misalignment of a fork end fitting can cause fretting fatigue problems. (This relates to Fig. 4.12.) Danger points indicated thus ■.

result in fretting and even anodised and painted surfaces may be abraded away as shown in Fig. 4.4. Nevertheless, modern paint primers developed for aircraft use have such a considerable resistance to abrasion that they play an anti-fretting role in resisting metal-to-metal contact. It is claimed that, with the exception of very severe cases, no specific anti-fretting compounds are needed to supplement this action.

Other features that present fretting hazards are shown in Fig. 4.5 and relate to washers or fastener heads that, either because of distortion in assembly, or because of incorrect geometry, do not seat uniformly. Another example, Fig. 4.6, will be recognised as a fairly common case where misalignment and/or side loads are applied in a fork fitting.

4.5.1 Practical Examples of Flange Type Fretting Fatigue

Figures 4.7 and 4.8 show cases of fretting fatigue in aluminium alloy aircraft skin stringer attachments where non-uniform closure conditions have existed and paint breakdown has occurred both at the skin/stringer interface and under fastener heads, and also where, by the design, local rubbing points exist at the edges of dimpled holes. There are special

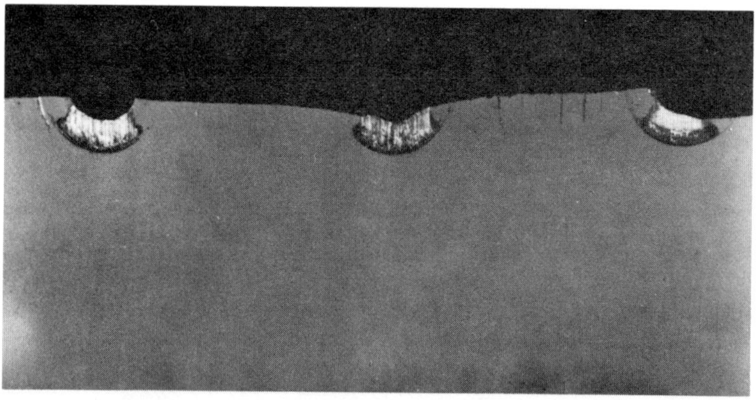

FIG. 4.7. Abrasion points on sheet arising from non-uniform closure loads under fastener heads. ×2·3.

problems experienced with aluminium alloys that have been clad with relatively soft aluminium (alclad). This cladding, being soft, is particularly prone to stage I fatigue cracking produced early in the life of the part. For this reason, introducing fretting, which also has the effect of producing early cracks, does not represent a particular fatigue penalty to these materials. Nevertheless, the presence of soft layers in a lap joint may

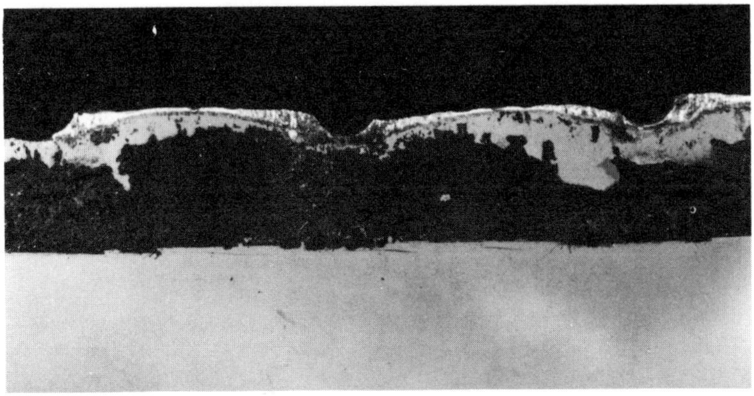

FIG. 4.8. Reverse face of sheet shown in Fig. 4.7. This face was painted and 'Thiokol' bonded but local closure pressure still led to metal to metal abrasion. ×2·3.

indirectly cause trouble by allowing relaxation of the normal fastener closure loads by creep, and wear under fastener end closures, and this, in turn, will encourage slippage. In addition, the softer coatings may be deformed more during drilling than unclad alloys and a larger burr will form. There have been many examples of loosening by wear under fastener

FIG. 4.9. Worn depression under fastener head leading to loosening of joint and final fretting fatigue from back face. ×2.

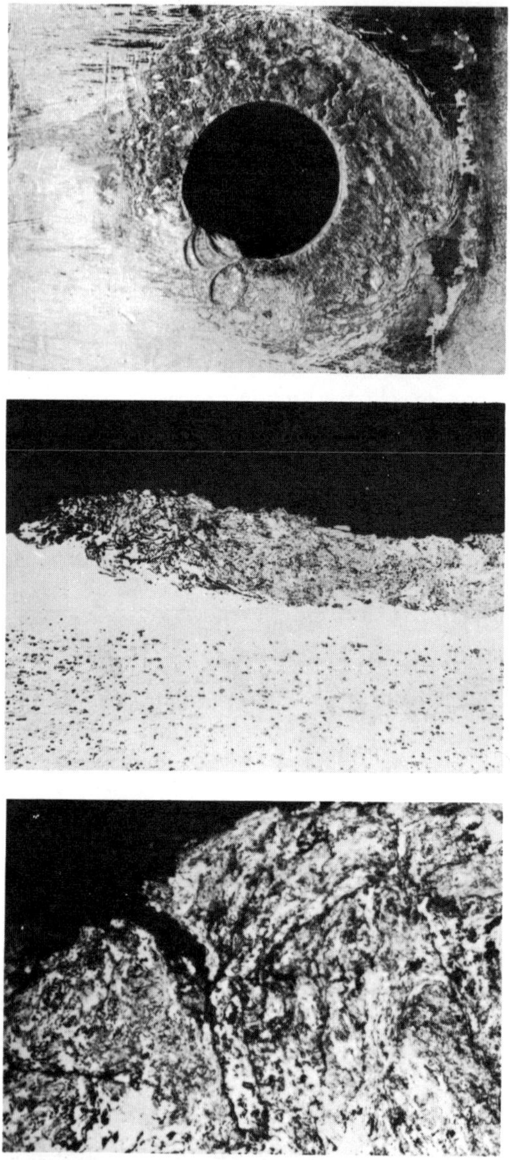

(a)

(b)

(c)

FIG. 4.10. Fretting damage on thick (5 mm) aluminium alloy alclad plate (BS L73). (a) Superficial appearance of fretting scar at fastener hole. × 3. (b) Section through fretting scar showing churned up zone containing a considerable amount of fragmented anodic coating. × 100. (c) Enlarged area of damage zone showing the development of cracks. × 500.

heads, one such example is shown in Fig. 4.9—in this case a stainless steel plate where fretting fatigue cracks have originated from the back face.

The past practice of proportioning the cladding thickness on alclad to that of the parent sheet or plate, commonly 5 % of the thickness per side, meant that for the thicker gauges the clad layer, being itself thicker, received less substrate support and quite excessive deformation was occasionally experienced. Figure 4.10 shows an example of this where the anodised layer was clearly fragmented and churned up by the heavy scuffing action that developed. It is now usual to limit the cladding thickness to 0·0125 mm per side irrespective of sheet or plate thickness.

Up to this point the examples of flange fretting that have been shown all relate to relatively thin sheet where manufacturing distortion is likely to occur. However, even in thick plate, local fretting may be experienced and,

FIG. 4.11. Fretting fatigue from contact with fastener heads where failure ignores the stress concentration of the holes. × 3.

as before, it will be related to the local closure pressures. Figure 4.11 shows an example where the fretting scar circumscribes the fastener hole and where, as is commonly experienced, the fatigue crack has been influenced more by the fretting than by the stress concentration of the hole itself.

The somewhat specific example, described earlier, of fretting between lugs and fork fittings is shown in Fig. 4.12. This is a 1-inch (25 mm) thick aluminium alloy member with a steel bushed hole. The fretting scar on the face of the member can be clearly seen, together with evidence of ingress of fretting product to the fatigue fracture face. The scar had developed from

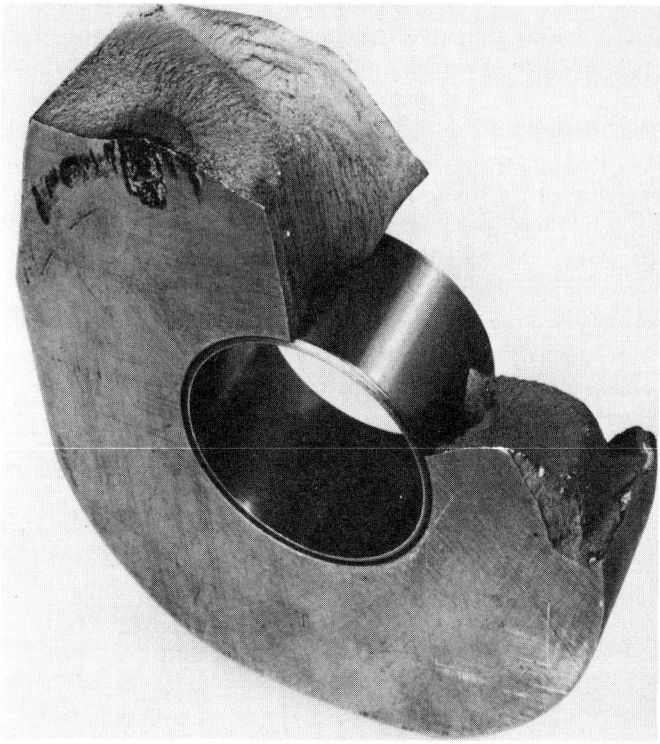

FIG. 4.12. Flange type fretting fatigue from interference point arising from misalignment of lug and fork fitting. Note that the fretting product had been forced into the crack.

interference between the member and the edge of the fork fitting into which it had been pinned.

As mentioned earlier, a considerable amount of fretting damage is bound to occur after fatigue cracking has allowed displacement of the material and in some cases it may not always be possible to apportion the blame on fretting. Figure 4.13 shows a well developed fretting scar that has formed, in the main, on one side of the crack. It is clear that most of this must have happened after the cracking event although it was virtually certain, in this case, that fretting had had some influence on the initiation of the crack.

The foregoing examples of fretting fatigue have been drawn from structural engineering practice where the loads experienced have been oscillatory. Even the fork fitting was part of a structural member, although a similar geometric form is frequently found on control rod ends. As would

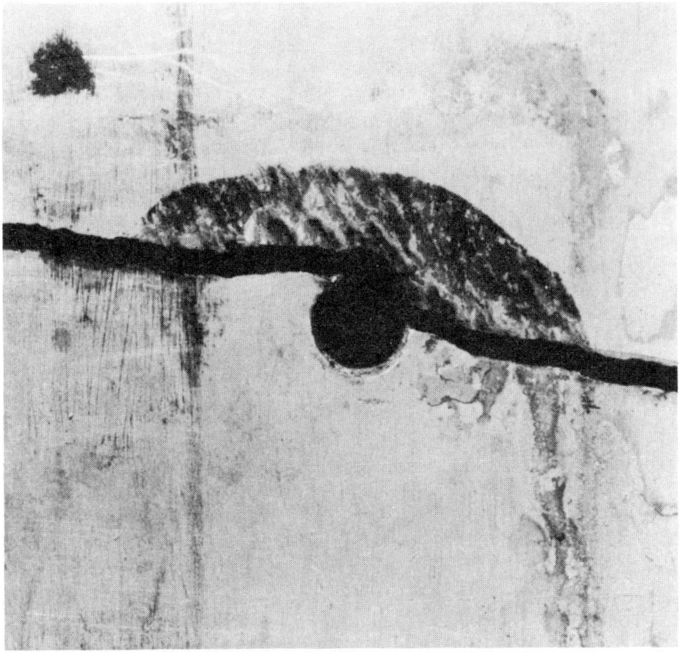

FIG. 4.13. A fretting scar that developed on one side of cracked tubular member (BS L63 aluminium alloy). ×2.

be expected, rotating machinery provides classic examples of fretting fatigue when the general definition of flange fretting rather than bearing fretting appertains. Keys and keyways can give trouble in this respect and the stresses developed in these positions may lead to torsional fatigue. Figure 4.14 shows the cross section of a steel shaft on to which was keyed a cast iron wheel. It can be seen that the fretting fatigue crack initiated, not at the point of stress concentration of the corner of the keyway, but somewhat higher up the side of the slot where the fretting scar had developed. This emphasises a point mentioned earlier, i.e. that fretting can override the geometric stress concentrations present, and an analysis of a number of cases has suggested that fretting has an equivalent effect to a notch with a theoretical stress concentration factor (K_t) equal to 4.

Splines on shafts are subjected to somewhat similar stressing circumstances as keyways, and the mode of initial failure can be as shown above, but the multiple cracks that frequently develop in a splined shaft may grow as torsional fatigue cracks to a state at which a considerable loss

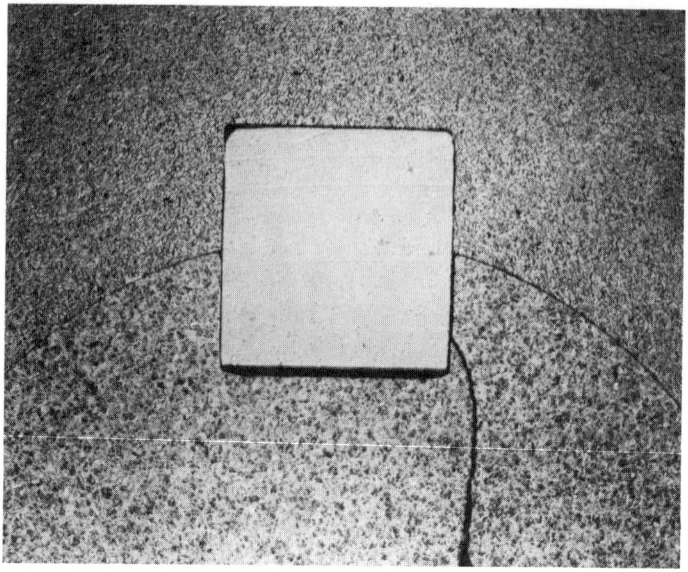

FIG. 4.14. Cross section of steel shaft with fretting fatigue crack developing from Keyway, and ignoring the stress concentration of the corner etched in Nital. ×7.

of torsional stiffness is experienced before they link with one another to completely separate the shaft.

Another classic problem related to rotating machinery occurs in the situation where a wheel is fitted to a shaft but where the interference is such as to allow some relative movement between the bore and the shaft surface during rotation.

4.6 PIN BORE FRETTING FATIGUE

In lap joints where fretting may lead to fatigue failure the crack will form in that member where the fatigue stress is highest, thus one part will act as a fretting pad to the other. In some instances it is perfectly possible for cracks to appear in both parts. The same situation can arise in pin and lug type fretting, but before describing examples it is necessary to explain why this type of fretting is presented as a somewhat different form to flange fretting. The interface between a pin or fastener and the bore of the hole in which it is situated clearly represents a situation where fretting might occur. In such a loaded hole the normal sliding loads are inter-related, whereas in the flange the normal loads obtain from the closure loads in the fasteners and are usually, although not always, independent of the sliding loads. The interdependence of the loads in the pin–bore assembly dictates the likely position of the fretting scar. Whereas fasteners are not expected to rotate, pins may well do so, which again differentiates this situation from that of

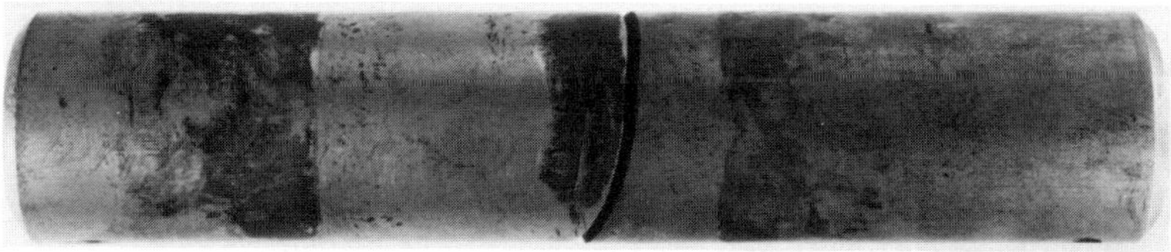

FIG. 4.15. Fretting fatigue failure of steel pin working against an aluminium alloy lug. ×1.

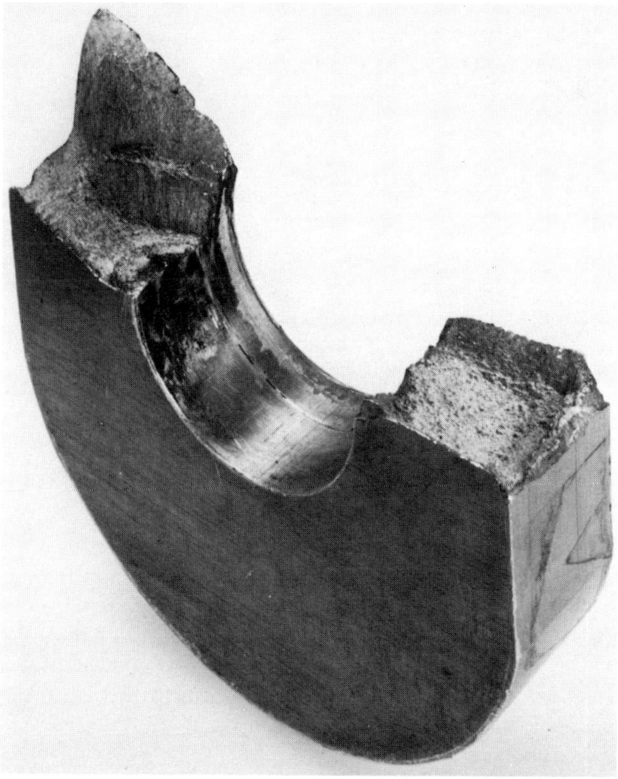

FIG. 4.16. Fretting fatigue failure in aluminium alloy (BS L65) lug. As was observed in Fig. 4.12 fretting product had been forced into the growing crack. ×1.

the flange. Again, fastener loads are usually applied in a fixed direction, albeit reciprocating, whereas pin–lug assemblies may be subjected to loads in various directions. In some cases pin failures may be experienced and in other cases, depending on the stresses present, fretting fatigue may occur in the lug or body in which the pin operates.

4.6.1 Practical Examples of Pin–Bore Failures

Figure 4.15 shows an example of a fretting fatigue failure of a 32 mm diameter steel pin where the fretting scars are related to the positions of load application in double shear whereas Fig. 4.16 shows a fretting fatigue failure in an aluminium alloy lug that resulted from fretting with the associated pin. Characteristically, some of the black fretting product of the aluminium alloy had been forced into the crack and can now be observed on the fracture surface in the vicinity of the crack origin. In the category of pin–bore failures can be included examples where failure is initiated in a countersink. As will be appreciated, sliding can occur between countersink screw heads and the countersink itself.

4.7 WIRE ROPES AND CONTROL CABLES

The assemblage of wires to form flexible cables presents a situation that is particularly prone to fretting. It is clear that flexure of the cable itself must result in the sliding of the individual wires against their neighbours and it is not uncommon to find that such abrasion produces noticeable flats in the areas of contact. When a cable runs over pulleys or fairleads, both the sliding and normal loads are increased and the motion of a cable to and fro constitutes a fatigue environment. Apart from mutual wear of neighbouring wires, wear can also occur against the pulley or fairlead surfaces themselves. It has long been recognised that the fatigue loading is dependent on the diameter of the pulley and the cross-section diameter of the cable itself, and certain limiting pulley diameters are specified in applications such as aircraft control cable systems, where cables are subjected to a large number of reciprocating actions. For corrosion protection it is common to use zinc coated wires but, more recently, stainless steel wires have also been brought into service. With their introduction came a spate of fretting fatigue cases. In retrospect, this does not seem surprising as the galling problem of stainless steel working against itself has long been recognised. It also seemed that zinc coating had been giving a measure of dry lubrication that had been beneficial. Nevertheless,

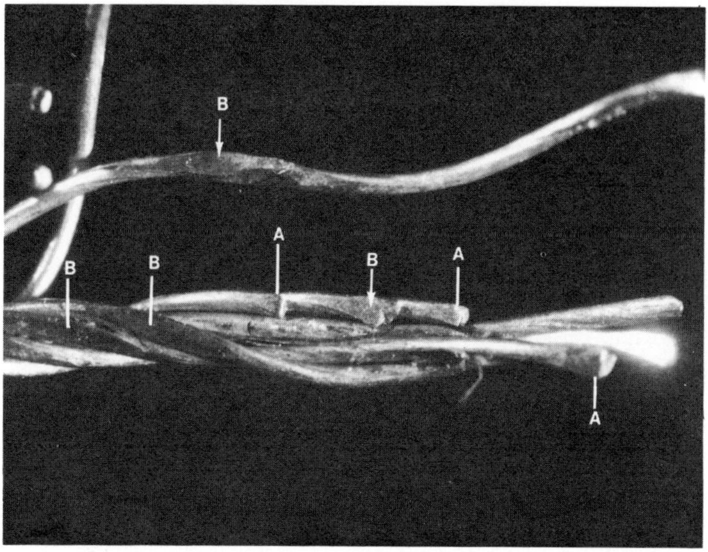

(a)

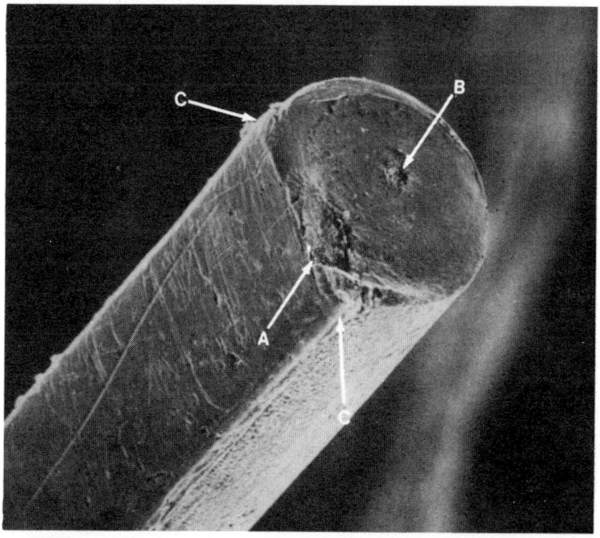

(b)

FIG. 4.17. Fretting fatigue failure of cable strands. (a) Unravelled strands showing fatigue fractures A and extensively worn flats B; (b) Fracture showing fatigue origin A, final torsional fracture point B, and limits of the worn flat C, C ($\times 200$).

stainless steel can give satisfactory service if the empirical rules regarding the relative diameters of cable and pulleys are obeyed. A characteristic of fretting fatigue of any cable is that the first failures frequently occur on those wires submerged near the centre of the cable and trouble may not be apparent until broken ends begin to protrude. This makes effective inspection virtually impossible, and policy dictates that control cables should be changed frequently, irrespective of superficial condition. Figure 4.17(a) shows an unravelled failed stainless steel control cable exhibiting fatigue fractures on individual strands (arrows A), and extensively worn flats (arrows B). Figure 4.17(b) illustrates an individual fracture with a fatigue origin (arrow A), and the final fracture point (arrow B) where torsional separation had occurred.

4.8 MORE UNUSUAL FRETTING FATIGUE CASES

It has been stated that the main fretting situations are generally appreciated and care is given, at the design stage, to guard against such events. Case histories of fretting fatigue failures, whether they lead to major accidents or only present a nuisance aspect to the operator, are frequently concerned with the more unusual and therefore the unexpected. These may be situations that would not immediately be recognised as potential hazards, even by experienced designers, or they may be cases that have arisen in areas of the structure not considered of prime importance, where the design work has been devolved to some less experienced individual.

The incidence of badly seated washers or fastener heads has already been mentioned. These would represent hazards outside the designer's control,

FIG. 4.18. Failed 'V' brace tube at plug end fitting. ×1.

being related mainly to the quality of the workmanship. Nevertheless, cases arise of failures where non-uniform closure is inherent in the design. Such a case is shown in Fig. 4.18 which illustrates the end fitting of a bracing member constructed as a tube in a high strength nickel chromium steel (BS 4T2). The end fitting itself was a cadmium plated steel part and the shear bushes were made in a medium strength high-chromium corrosion-resisting steel (BSS 80). Figure 4.19 shows schematically the detail of this assembly and Fig. 4.20 shows the fatigue fracture that had developed from the point of contact with the shear bush.

An examination of this fracture in the tube revealed that there was a fretting fatigue origin which coincided with a patch of fretting on the underside of the shear bush head, shown in Fig. 4.21. Although the holes had been carefully reamed to accept the shear bushes it is clear from considerations of the curvature of the tube itself that a reasonably large

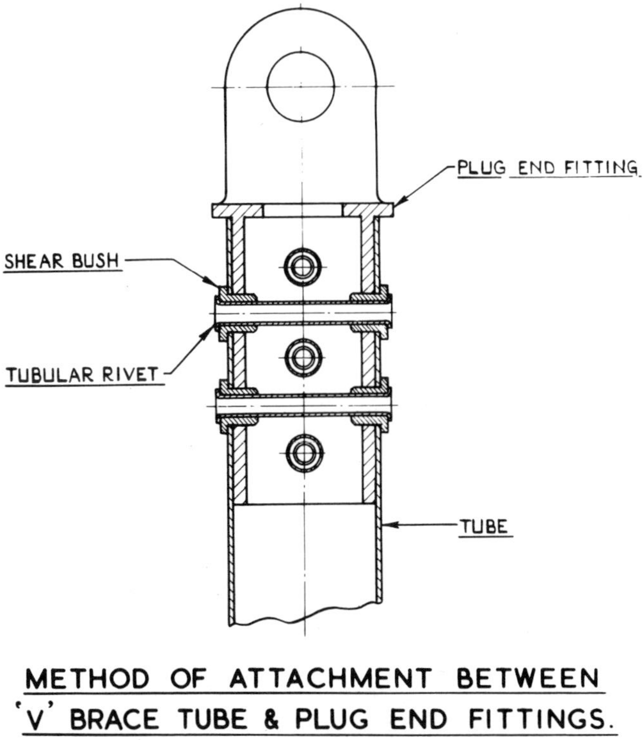

FIG. 4.19. Schematic diagram showing the detail of the attachment between tube and plug end fitting.

FIG. 4.20. Fretting fatigue origin of tube at point of contact with shear bush.

FIG. 4.21. Associated shear bush showing fretting scar in fillet.

seating area could not be achieved, and it might have been anticipated that wear would occur at the relatively small contact zones, as in fact happened.

It is often difficult in a potentially dangerous fatigue situation to entirely avoid fretting. Most of the classic situations, particularly the fastened flange joint, can be satisfactorily saved from fretting, but some engineering arrangements are inherently dangerous and difficult to control. One such case is the aircraft lift strut stabilising point where aerodynamic vibrations are developed in the member. The strut itself was constructed from a BS L63 aluminium alloy hollow extrusion. Over a number of years several cases of fretting fatigue have been experienced in the region of the stabiliser strut attachment assembly, although these cases differed inasmuch as the fretting points were not always at the same position in the assembly. The lift strut attachment point is shown in Fig. 4.22, and a schematic drawing, Fig.

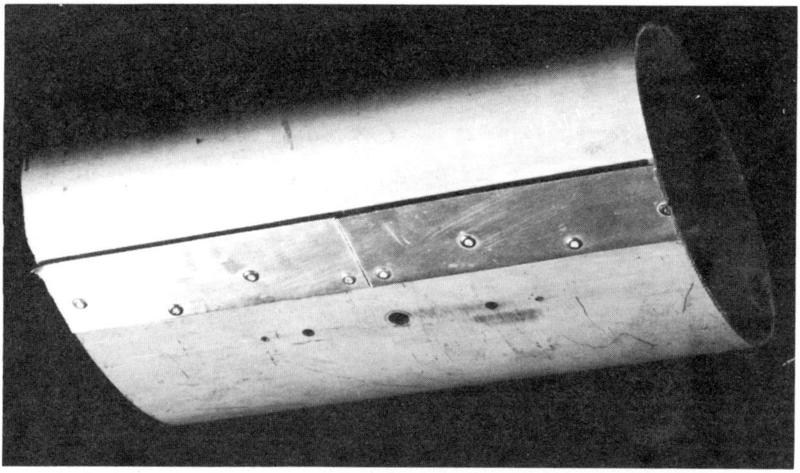

FIG. 4.22. Aeroplane lift strut. Stabilising strut attachment position showing lower surface of lift strut where cracks were discovered at bolt hole.

4.23, shows the detail. It is clear from this arrangement that flexure of the tubular strut will cause a shape change in the tube section resulting in sliding or hammering actions at several points. Where these points coincided with regions of fatigue stress, growing fatigue cracks were experienced. One such point was on the lower surface of the hollow spar at the hole where the washer on the attachment bolt made contact. Figure 4.24 shows the lenticular fretting scar A, and the fatigue crack that had developed, B. Figure 4.25 shows a view of this crack after opening, together with another

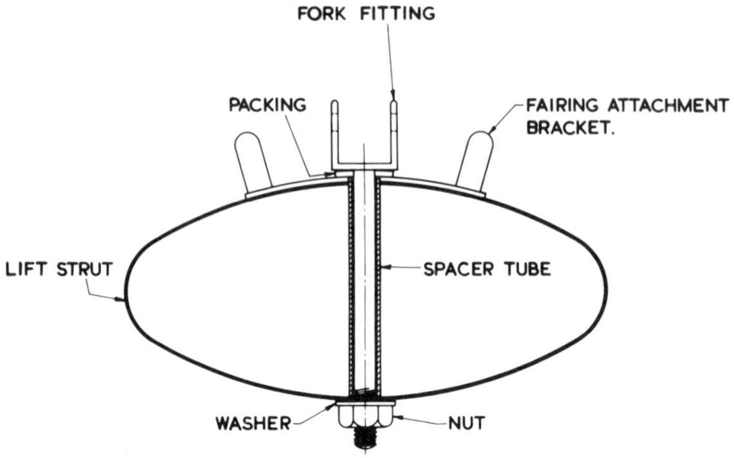

FIG. 4.23. Schematic diagram showing the detail of the method of attachment of the stabilising strut (not to scale).

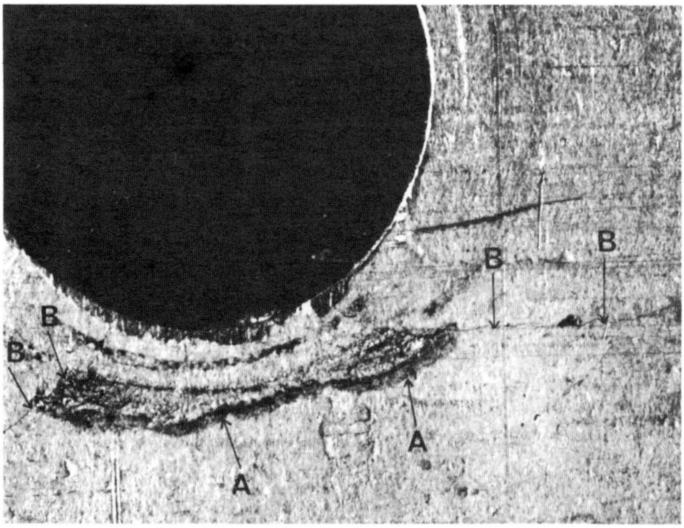

FIG. 4.24. Fretting fatigue scar produced by contact with aluminium alloy washer, arrows A, and fatigue cracks developing from fretting scar, arrows B.

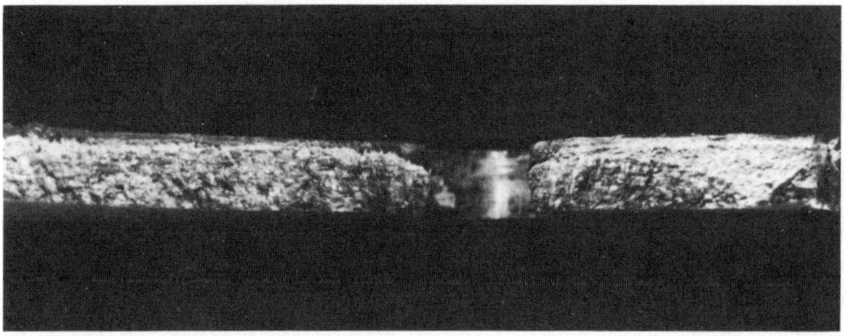

FIG. 4.25. Cracks shown in Fig. 4.24 after opening. Note that the origins are at the fretted surface and the stress concentration of the fastener hole itself has had no influence on the failure.

crack that had developed from the other end of the fretting scar. In another instance of a fretting fatigue failure in this type of structure, fatigue cracks had developed at the top hole in the strut as shown in Figs. 4.26 and 4.27. In this case fretting scars had developed in the bore from contact with the spacer tube, where an associated scar was observed, Fig. 4.28. It is evident that there was a design fault present inasmuch as there was no positive

FIG. 4.26. Lift strut of a similar detail design to that illustrated in Fig. 4.22, but where, in this case, fretting fatigue cracks have appeared at the upper surface hole. ×1.

FIG. 4.27. Detail of cracks relating to Fig. 4.26 showing that the fretting scars, arrow A, in this instance, are in the bore of the hole in which the end of the spacer tube is located. ×4.

spacing action provided, the tube being free to slide with respect to the tubular strut wall. This fault was later rectified by a redesign in which a considerably stouter internal spacer block was introduced. This case presents a classic example of failure to identify the various sliding actions that can occur in a somewhat unusual assembly.

There are many other examples of fretting fatigue failure where relative slip of contact faces is inherent in the design. The possibility of such movements is not always recognised at the design stage or the dangers of fretting fatigue may not have been appreciated. Figure 4.29 shows a failed part of a flexible coupling used in a drive shaft assembly in which flexibility is achieved by using thin leaves of stainless steel sandwiched together by fasteners. The danger of stainless steel working against itself was mentioned in the context of wire ropes, and again in this instance trouble was experienced. Figure 4.30 shows the fretting fatigue fracture in more detail and Fig. 4.31 is a microsection through the fretting scar where many other fretting fatigue origins can be seen.

A further case where fretting fatigue failures were experienced, and again, an example that cannot be simply categorised into flange type or loaded

FIG. 4.28. Associated fretting scar, arrow A, on spacer tube. ×2.

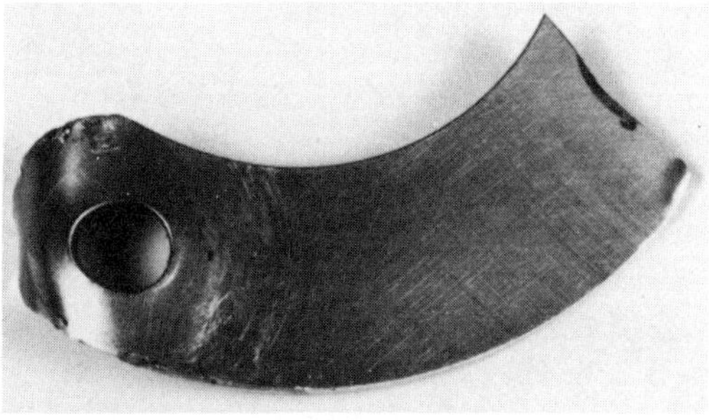

FIG. 4.29. Failed leaf of a flexible coupling from a drive shaft of 18/8 stainless steel.

FIG. 4.30. Fretting fatigue fracture related to Fig. 4.29.

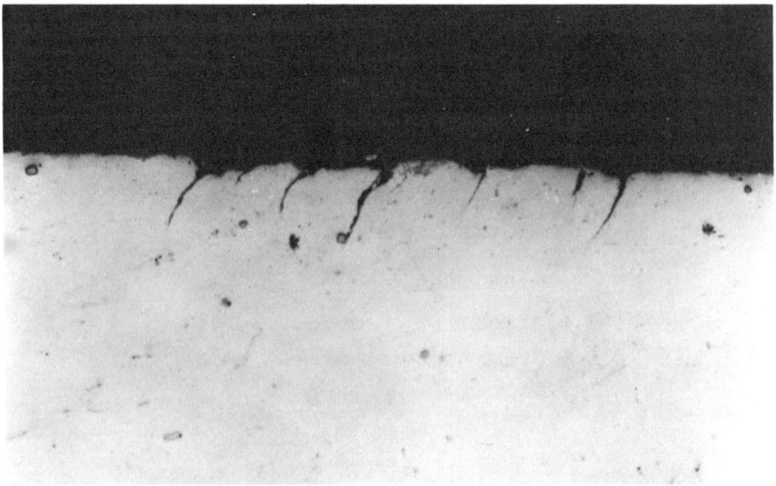

FIG. 4.31. Microsection related to Fig. 4.30 showing associated secondary fretting fatigue origins. ×250.

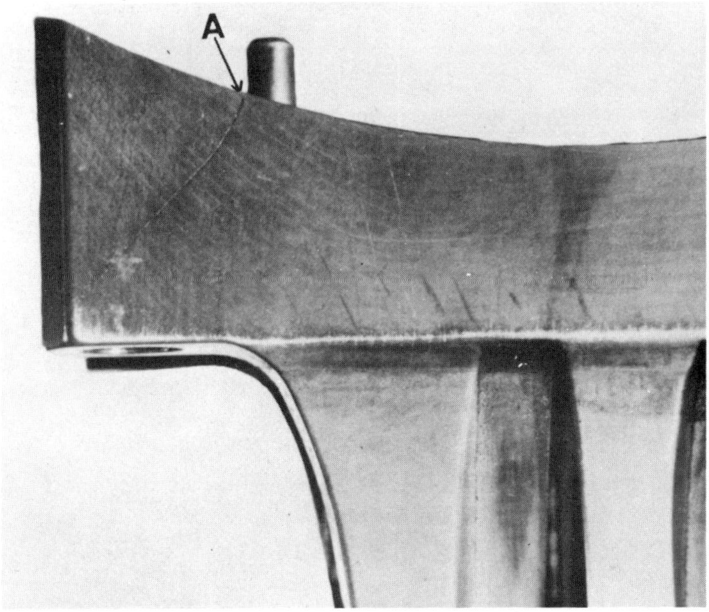

FIG. 4.32. 'Big end' palm cracking (Diesel engine)—crack shown at A.

FIG. 4.33. Fretting scar related to Fig. 4.32—crack shown at A.

FIG. 4.34. Fracture surfaces after opening crack showing in Fig. 4.33—fatigue origin indicated by arrow A.

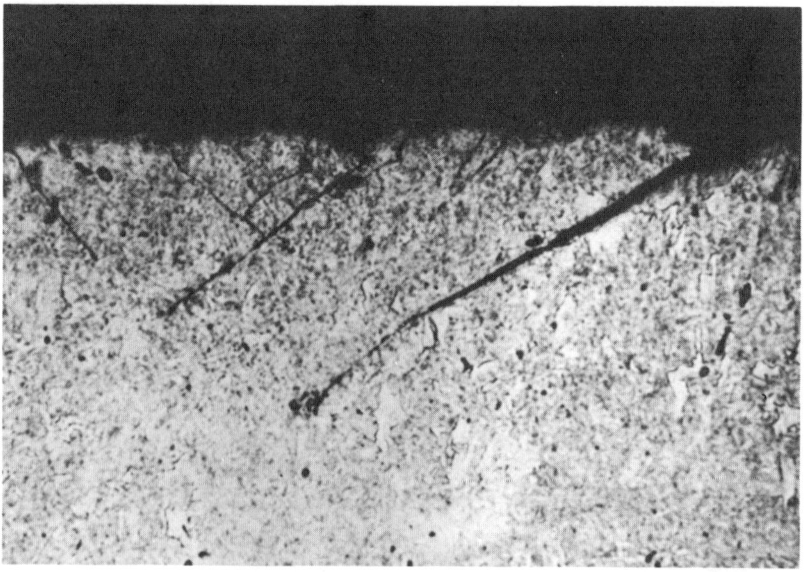

FIG. 4.35. Microsection through failed region showing typical stage I secondary fretting fatigue cracks. ×500.

hole type fretting is the 'big end' palm from a diesel engine. The connecting rod loads are distributed over this palm and thus transmitted to the bearing. Cracks were experienced at the position shown in Fig. 4.32, and Fig. 4.33 shows the fretting scars associated with cracking. The origin was at the position indicated by the arrows in Fig. 4.34, which shows the opened fracture. A microsection through the line shown X–X in Fig. 4.33 revealed the presence of other typical fretting fatigue cracks, as can be seen in Fig. 4.35. The palm face was 'as machined' steel, whilst the bearing housing face was chromium plated. The relative sliding motion arose from a degree of flexure resulting from the offset compression loads on the face of the palm, and an unsuccessful attempt had been made to reduce these by introducing the peg shown in Fig. 4.32. A further attempt to remove the problem was made by introducing a lead coating on the palm face but, again, with no success. The problem was finally solved by nitriding the palm face.

4.9 GENERAL CONCLUSIONS AND COMMENTS

The danger with fretting, as with corrosion, is the devastating effect that it has on total fatigue life by virtually eliminating the crack initiation period, which in many cases represents most of the predicted life of the component. In relatively low cycle fatigue designs it is usual to apply a factor to the life as determined by the fatigue test programme (which may well be on one specimen). For example, the safe aircraft structure life may be assessed as being either 1/3 or 1/5 of the life sustained by the test. Nevertheless, it is clear that in spite of this seemingly generous factor, if fretting conditions have not arisen at the fatigue critical positions in the test specimen (perhaps because of some of the variables of fit and assembly described earlier) and yet they may arise in service, then there is a strong possibility of a premature failure.

The evidence is that once a fretting fatigue failure has occurred in service, remedial measures can be found which will eliminate the problem. Some cases are remarkably intractable and may require a change in both geometry and material. It is clear that an improved understanding of the contribution of the various factors involved has led to a general diminution of the incidence of fretting fatigue failures in such recognised danger areas as fastened flange joints. Nevertheless, difficulties still arise in applications where the possibilities of fretting have not been identified at the design stage, and in pin–lug combinations where the engineering variables are difficult to control and where anti-fretting compounds are difficult to maintain *in situ*.

5

Practical Observations of Initiation and Propagation of Fretting Fatigue Cracks

K. Endo

Department of Mechanical Engineering, Kyoto University, Kyoto, Japan

5.1 INITIATION OF FRETTING FATIGUE CRACKS

The definition of fatigue crack initiation is difficult. This is especially so under fretting fatigue where specimen surfaces are severely damaged and slip lines or micro-cracks are rarely observed. Nevertheless, the very early initiation of visible cracks is a characteristic of fretting fatigue. The subsequent crack growth is very slow because of the low applied stress. The maximum frictional stresses on the fretting surface are repeated at the edges of the contact region and cracks are generated near this area. When the fretting amplitude is so small that no macro-slip occurs, cracks are generated at the boundary between the slip and the non-slip in the contact region,[1] as is known from the stress distribution. When repeated stresses are applied in the same phase with repeated frictional stresses, i.e. under fretting fatigue, crack initiation is accelerated and the crack propagates within the material and results in fracture. This chapter is based on research on initiation and propagation of fretting fatigue cracks[2] and effects of environment on them.[3]

Figure 5.1 shows the damage curves of fretting fatigue of an annealed 0·34%C carbon steel (yield strength 280 MN m^{-2}, ultimate tensile strength 550 MN m^{-2}, Vickers hardness 135 kgf mm^{-2}, endurance limit in air 210 MN m^{-2}). The plate test piece, to the central surface of which is clamped a cylindrical pad 12 mm in diameter, made of the same materials, under line contact pressure of $p = 39$ kN m^{-1} is subjected to reversed bending of 30 Hz. After n_f cycles of fretting fatigue, the fatigue test is continued without fretting and the total life N_t is obtained. In the figure, the broken line representing $n_f/N_t = 100\%$ shows the failures by continuous fretting fatigue throughout the life. The fretting fatigue life at a repeated stress of $\sigma_0 = 190$ MN m^{-2} is about $N_t = 4 \times 10^5$ in air and $N_t = 10^6$ in

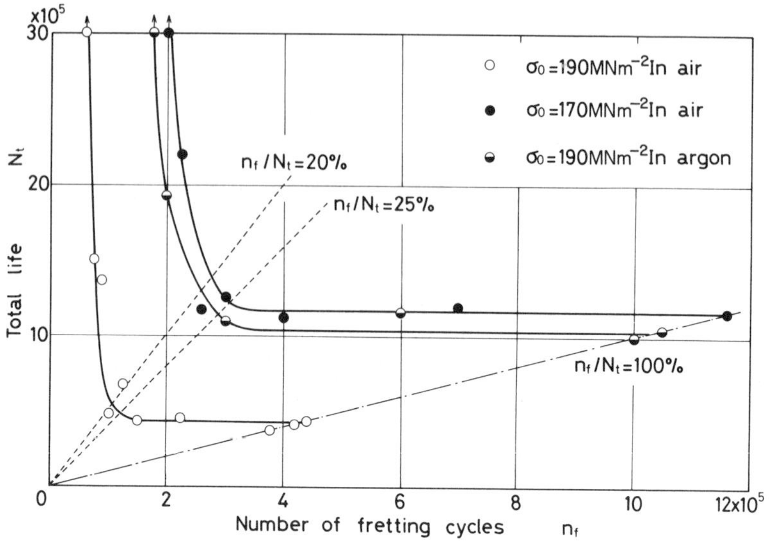

FIG. 5.1. Fretting fatigue damage curves of 0·34% C carbon steel under reversed bending in air and in argon ($p = 39\,\mathrm{kN\,m^{-1}}$, $f = 30\,\mathrm{Hz}$).

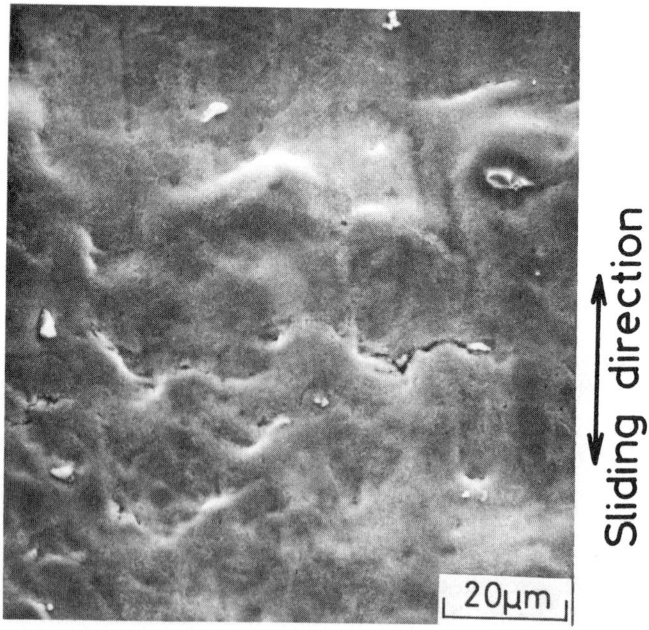

FIG. 5.2. Fretting surface of a 0·34% C carbon steel at an extremely early period ($n_f = 10^4$).

argon. Fretting fatigue damage is known to saturate at $n_f/N_t = 20$–25%. The fretting fatigue limit defined by Wharton *et al.*,[4] i.e. the critical number of cycles showing no fretting damage, is also apparent.

The surfaces of specimens tested for fretting fatigue at $\sigma_0 = 190\ \mathrm{MN\,m^{-2}}$ were lightly electropolished and were observed in a scanning electron microscope. The slip amplitude was 20 μm at steady state. Figure 5.2 shows the fretted surface at a very early stage, i.e., $n_f = 10^4$ ($n_f/N_t = 0{\cdot}025$); a few micro-cracks of 20–30 μm in length can already be seen. At $n_f = 4 \times 10^4$ ($n_f/N_t = 0{\cdot}10$) many cracks of 20–70 μm in length are observed at the edge of the contact area (Fig. 5.3). In this stage the cracks

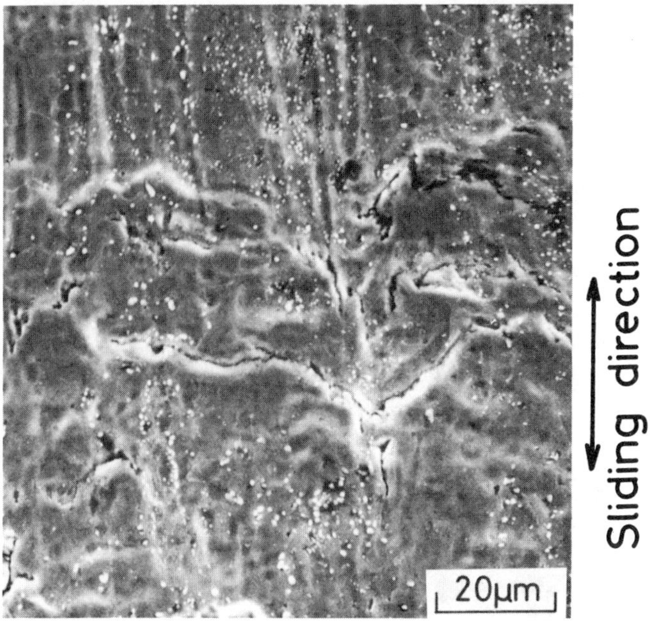

FIG. 5.3. Small fatigue cracks near the end of contact ($n_f = 4 \times 10^4$).

are smaller than 10 μm in depth, and under some circumstances may disappear due to fretting wear. The cycle number of $n_f = 4 \times 10^4$ is still below the fretting fatigue limit as is seen in Fig. 5.1. Figure 5·4 is a fretted surface at $n_f = 10^5$ ($n_f/N_t = 0{\cdot}25$) which is etched to show the microstructure. Cracks of a length of several hundreds of microns are propagating perpendicularly to the sliding direction of fretting and are seen traversing the pearlite colonies. Where $n_f/N_t = 0{\cdot}25$ is the stage of saturation of the

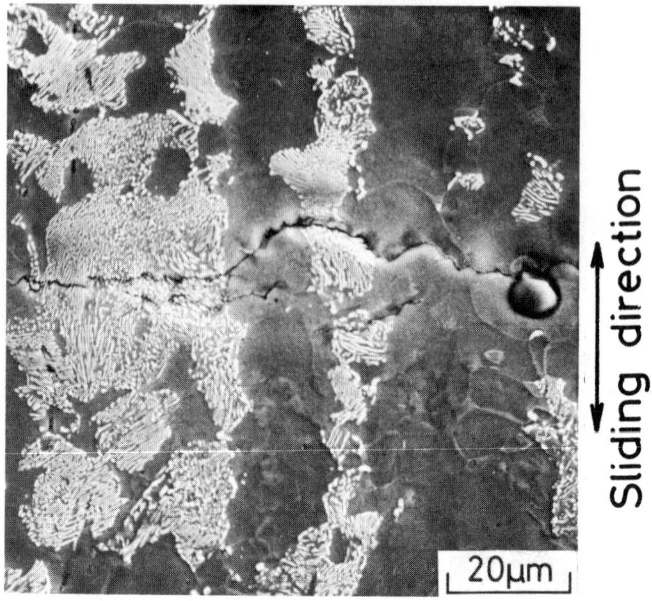

FIG. 5.4. A large crack on the contact surface ($n_f = 10^5$).

fretting damage in Fig. 5.1, and since the contact width increases thereafter due to wear, cracks are located a short distance inside the edge of the contact region. These cracks gradually become interconnected and propagate into the material as a macro-crack resulting in fracture.

The residual stress in the fretting surface was measured by X-ray methods

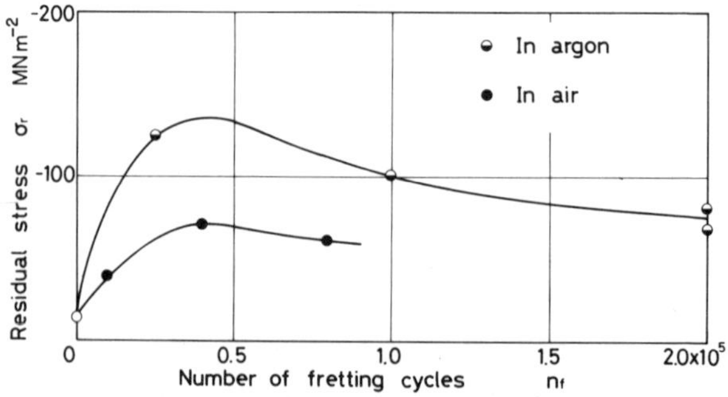

FIG. 5.5. Changes of residual stress on the contact surface of specimens of 0·34 % C carbon steel against the fretting cycles in air and in argon ($\sigma_0 = 190\,MN\,m^{-3}$, $p = 39\,kN\,m^{-1}$, $f = 30$ Hz).

in the sliding direction, i.e. in the direction of repeated stress. As seen in Fig. 5.5, the compressive residual stress increases initially and then gradually decreases after showing a peak. The cycle number at the peak corresponds approximately to $n_f = 4 \times 10^4$ ($n_f/N_t = 0 \cdot 1$) where many micro-cracks are initiated in air.

The variation of coefficient of tangential force with the number of fretting cycles is shown in Fig. 5.6. There is no difference in the variation of

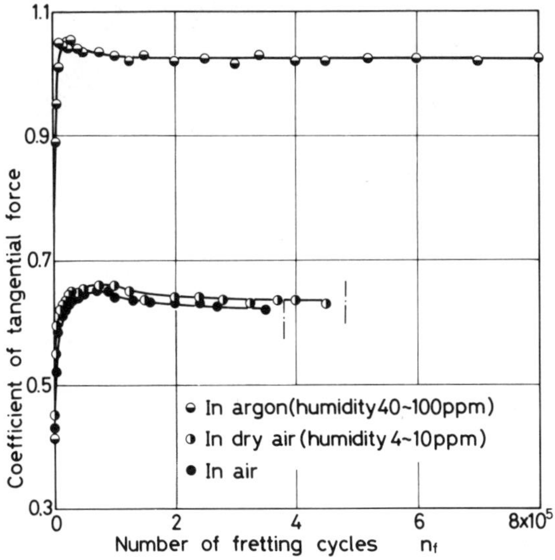

FIG. 5.6. Variation of coefficient of tangential force of 0·34% C carbon steel in various environments ($\sigma_0 = 190$ MN m^{-2}, $p = 39$ kN m^{-1}, $f = 30$ Hz).

tangential force in room air (humidity 10^4–2×10^4 ppm) and in dry air (humidity 4–10 ppm). The effect of humidity on the tangential force in air is negligible. The fretting fatigue strength at $N = 10^7$ in both the environments was 150 MN m^{-2}. The tangential force in argon (humidity 40–100 ppm) is about 1·7 times that in air and the wear debris consists of a metallic powder. The fretting fatigue strength in argon was 175 MN m^{-2}.

The Vickers hardness of ferrite crystals on the surface measured after slight electropolishing was about 180 when fretted in room air and about 300 in argon. Work hardening by adhesion is known to be more severe in argon than in air. This explains the higher residual stress in argon shown in Fig. 5.5.

Figure 5.7 is the fretted surface at $n_f = 5 \times 10^4$ tested in argon under σ_0

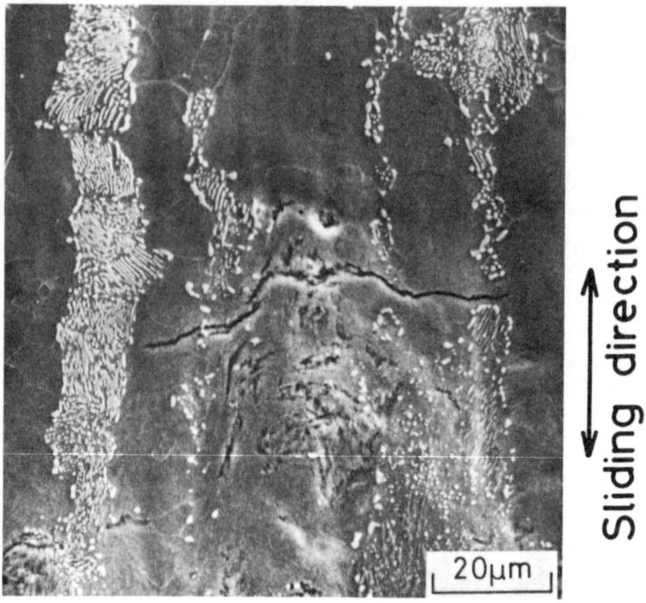

FIG. 5.7. Small fatigue crack at $n_f = 5 \times 10^4$ tested in argon ($\sigma_0 = 190\,\text{MN m}^{-2}$, $p = 39\,\text{kN m}^{-1}$, $f = 30$ Hz).

$= 190\,\text{MN m}^{-2}$, $p = 39\,\text{kN m}^{-1}$ and $f = 30$ Hz, and electropolished lightly to show microstructures. Although surface oxides are little in evidence, many micro-cracks are initiated at a similar number of cycles as in air (cf. Fig. 5.3). This may be due to the opposing effects of the increase in tangential force mentioned above and the deceleration of crack initiation due to the reduced aggression of the oxygen-free environment. The difference in fatigue life is due to the crack propagation rate being affected by the presence of oxygen as will be seen later in Fig. 5.14. However, the effect of humidity is rather small in such a ductile carbon steel.

Aluminium alloys are known to be very sensitive to water corrosion under dynamic conditions. The fretting fatigue tests on Al–4Zn–2Mg–T6 (yield strength $410\,\text{MN m}^{-2}$, ultimate tensile strength $450\,\text{MN m}^{-2}$, Vickers hardness $91\,\text{kgf mm}^{-2}$, fatigue strength at $N = 10^7$ in air $140\,\text{MN m}^{-2}$) were carried out in the same way as on the carbon steel. The line contact pressure was $8\,\text{kN m}^{-1}$ and the frequency was 20 Hz. The fretting fatigue strength at $N = 10^7$ in argon (humidity 40–100 ppm) was $116\,\text{MN m}^{-2}$ and was higher than $103\,\text{MN m}^{-2}$ in room air (humidity 10^4–2×10^4 ppm), lower than $130\,\text{MN m}^{-2}$ in dry air (humidity

4–10 ppm), and nearly equal to that in dry air (humidity 40–100 ppm). The tangential force in these environments was 32, 66 and 65 N respectively. The humidity decreases the tangential force but the extent is rather small. The removal of oxygen decreases the tangential force considerably because of soft metallic wear debris accumulating between the mating surfaces. Crack initiation by fretting fatigue in aluminium alloys is, however, accelerated by traces of water vapour due to corrosive attack but not by oxygen, as will be deduced from Fig. 5.15.

5.2 PROPAGATION OF FRETTING FATIGUE CRACKS

The combination of frictional stress and repeated stress accelerates crack propagation as well as crack initiation. Figure 5.8 shows the propagation of a fretting fatigue crack in a specimen of 0·34 %C carbon steel under reversed

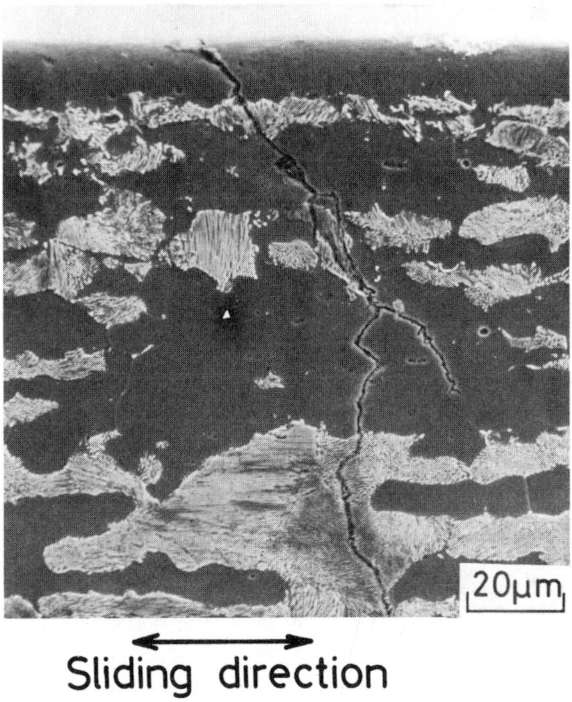

FIG. 5.8. Fretting fatigue crack propagation into a specimen of 0·34 % C carbon steel under reversed bending of $\sigma_0 = 190\ \mathrm{MN\,m^{-2}}$, $p = 39\ \mathrm{kN\,m^{-1}}$ and $f = 30$ Hz.

bending. The shear type crack of stage I[5] which is seen in the very early stages of a normal fatigue crack is not apparent in fretting fatigue because of severe cold working and wear of the surface layer. The tensile type crack of stage II which is perpendicular to the principal tensile stress propagates obliquely to the surface due to the tangential stress combined with the repeated direct stress. When the crack depth becomes larger, the direction of crack propagation becomes perpendicular to the direction of the axis of the specimen as the effect of the tangential stress on the contact surface is eliminated. The turning depth is about 70 μm under these test conditions.

The crack depth, l, was obtained by optical microscope and electrical potential technique. Multiple cracks were found by the optical microscope on both sides of the rectangular specimen. The maximum crack depths perpendicular to the fretting surface were measured at each fretted area and the mean value was determined as a crack depth l. The electrical potential technique was applied by measuring continuously the change in electrical resistance between two points at both areas of fretting contact which was due to constriction of electric current flow by the cracks. A linear relation was obtained between crack depth in a direction perpendicular to the surface and change in electrical potential. The maximum crack depth was derived from the electrical potential change using the linear relation. The relationships between fretting cycle n_f and crack length l shown in logarithmic scale are illustrated in Figs. 5.9 and 5.10, in which the vertical broken lines indicate cycles to fracture. In the figures, the relation $\log l = cn$

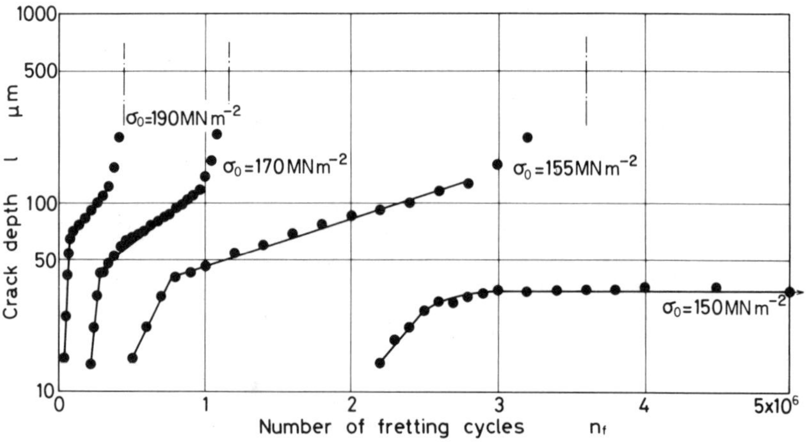

FIG. 5.9. Fatigue crack propagation curves of 0·34% C carbon steel for various stress amplitudes ($p = 39$ kN m^{-1}, $f = 30$ Hz).

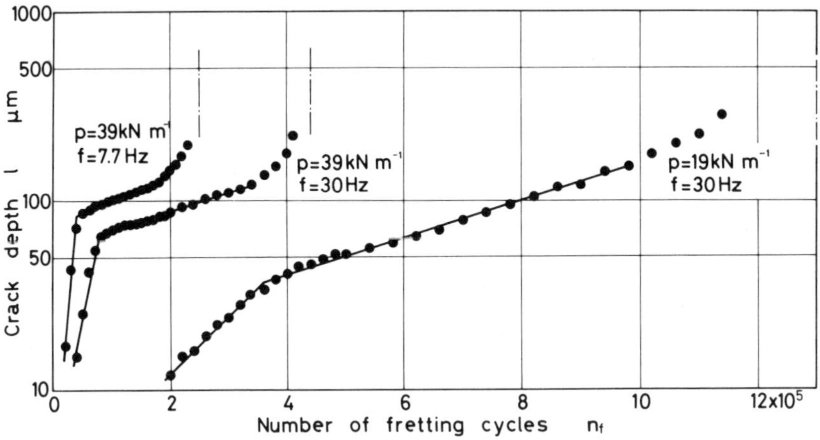

FIG. 5.10. Fatigue crack propagation curves of 0·34% C carbon steel for various contact pressures and frequencies (σ_0 = 190 MN m^{-2}).

is observed where c is a proportional constant. In other words, the crack rate dl/dn_f is proportional to the crack depth, l. The propagation curve, however, consists of two straight lines, showing two values of c, except for the final period of life. The number of cycles at which a crack of 10–20 μm in depth is detected is smaller at a higher repeated direct stress σ_0 or where the tangential force is higher (higher contact pressure p, lower frequency f). This number of cycles corresponds to the fretting fatigue limit in Fig. 5.1. A fatigue crack larger than this is understood as a propagating crack.

Below a given depth, the effect of tangential force becomes negligible. At the knee in the curve the crack propagation is represented by another straight line. The gradient of each crack propagation curve above the knee is the same under a constant repeated stress σ_0, in spite of different contact pressure p and frequency f as is seen in Fig. 5.10.

The crack rate dl/dn_f is reduced under lower repeated stresses as is seen in Fig. 5.9. At σ_0 = 150 MN m^{-2} the crack remains dormant at 35 μm in depth. The non-propagating crack is usually seen in sharply notched specimens. Under fretting fatigue, however, smooth specimens frequently have non-propagating cracks and this may be due to the compressive residual stress in the surface layer (Fig. 5.5) as well as the effect of the tangential stress decreasing with depth. Another example of non-propagating cracks is shown in Fig. 5.11.[6] A normalised 0·37%C carbon steel (Vickers hardness 170 kgf mm^{-2}) was subjected to plane-bending with fretting against cylindrical pads of 15 mm in radius made of a high carbon

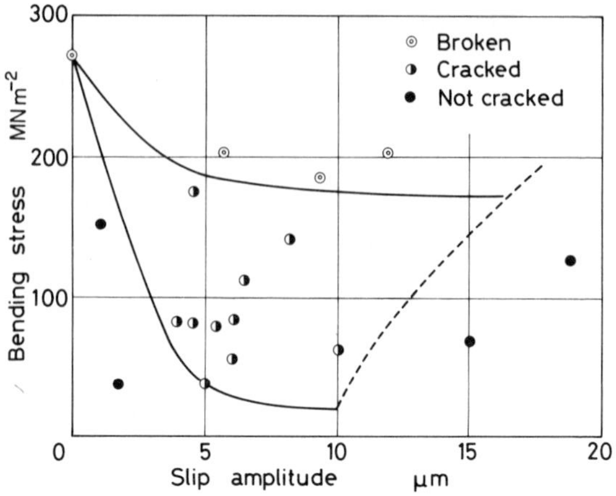

FIG. 5.11. Fretting fatigue strength at $N = 10^7$ of a normalised 0·37 % C carbon steel under $p_0 = 530\ \mathrm{MN\,m^{-2}}$.

steel (Vickers hardness 250 kgf mm^{-2}). The contact pressure was 530 MN m^{-2} in Hertzian stress. In the figure, the bending stress initiating cracks reduces with the slip amplitude to a few tens of MN m^{-2} and remains nearly constant at slip amplitudes of about 10 μm. The bending stress which causes failure due to propagating cracks is, however, 180 MN m^{-2}, being considerably higher than the crack initiation limit. Thus a wide range of bending stresses between 180 MN m^{-2} and a few tens of MN m^{-2} results in non-propagating cracks in medium carbon steels.

The depth corresponding to the knee at $\sigma_0 = 190\ \mathrm{MN\,m^{-2}}$, $p = 39\ \mathrm{kN\,m^{-1}}$ and $f = 30$ Hz is about 70 μm which agrees well with the change in direction of the crack shown in Fig. 5.8. The depth of the knee is greater when the tangential force is greater (p is greater and f is smaller) as is seen in Fig. 5.10 and when the applied repeated stress is greater as is seen in Fig. 5.9. This is because the effect of the combined stress extends more deeply within the material. Since the crack propagation curves above the knee are parallel with each other under different fretting conditions in Fig. 5.10, i.e., the crack rate dl/dn_f is the same at the same crack depth l, the crack propagation corresponds to that of a fatigue crack in a pre-cracked specimen subjected only to the repeated stress σ_0.

Crack propagation in fretting fatigue is very similar to that in corrosion fatigue but their mechanisms are different. The crack propagation curve in corrosion fatigue also consists of two straight lines at frequencies higher than about 0·1 Hz.[7] The reason is the reduction of stress amplitude at the crack tip due to the wedge effect of viscous corrosion products on the crack wall or, in other words, the reduction of the range of effective stress intensity factor due to the enhanced crack opening stress by corrosion products when the crack depth exceeds a given value. The main reason for decreased crack rate at higher frequencies is retardation of changes in corrosion potential with stress variation,[8] but the stronger wedging action is another reason at higher frequencies. The crack growth rate in fretting fatigue decreases at higher frequencies in the same manner as corrosion fatigue, but the main reason is the reduced frictional force resulting from accumulation of oxide debris between the mating surfaces.

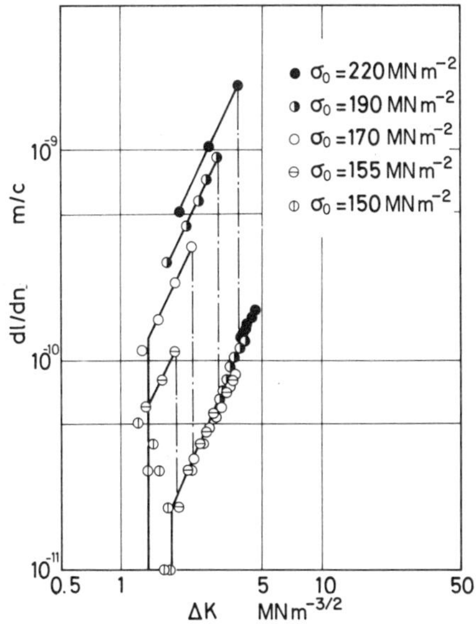

FIG. 5.12. Graph of dl/dn_f v. ΔK in fretting fatigue of 0·34% C carbon steel under various stress amplitudes ($p = 39\,kN\,m^{-1}$, $f = 30$ Hz).

The crack rate dl/dn_f is calculated using the straight lines of the crack propagation curves in Figs. 5.9 and 5.10 and is plotted in Figs. 5.12 and 5.13 against the range of the stress intensity factor ΔK. In the figures, the stress ratio is $R = K_{min}/K_{max} = -1$, and K is $\Delta K = K_{max}$, K_{min} being taken to be zero in accordance with ASTM E647. The calculation of K is made for pure bending of an edge-notched rectangular beam. The K-value in linear fracture mechanics should be applied to the large crack length above the knee point in Figs. 5.9 and 5.10, but is applied correspondingly to the small

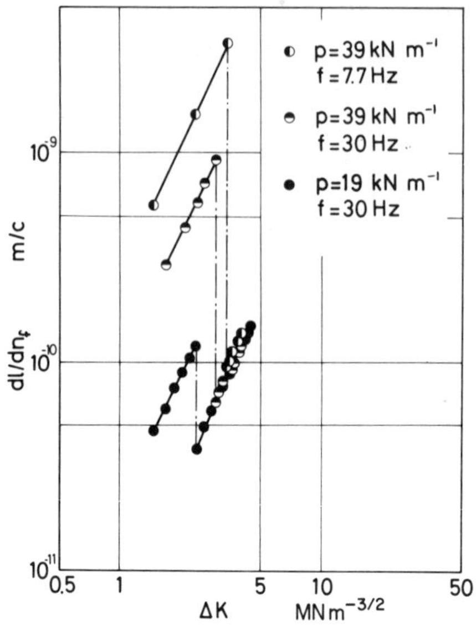

FIG. 5.13. Graph of dl/dn_f v. ΔK in fretting fatigue of 0·34% C carbon steel under various contact pressures and frequencies ($\sigma_0 = 190\ MN\,m^{-2}$).

crack length below the knee point to determine the transition phenomena. Furthermore, though K should include the mode II crack (shear mode), the K values in the figures are calculated on mode I crack (opening mode) by considering only the component of repeated bending stress. This is because of the varying effect of surface friction on the stress intensity factor as the crack propagates and this treatment may make more clear the effect of tangential force on the crack rate.

In fretting fatigue too, the relation

$$\frac{dl}{dn_f} = C(\Delta K)^m \tag{5.1}$$

is found, and the upper group of straight lines passes to the lower group showing little transition region. Hoeppner is of the opinion that a transition region is to be recognised.[9] As the presentation on a logarithmic scale, however, hardly shows the transition region, vertical broken lines are considered appropriate for the transitions.

The upper group of straight lines in Figs. 5.12 and 5.13 shows the crack propagation in the earlier range where the frictional force is effective. The value of C in the above equation is dependent on contact pressure p as well as on repeated stress σ_0 i.e. slip amplitude. The value of m seems to be constant. On the other hand the lower group of straight lines is the crack rate above the knee of the crack propagation curves in Figs. 5.9 and 5.10 and is represented by a single straight line whose gradient m is equal to 2. The crack rate depends only on ΔK calculated solely from the repeated bending stress. There seems to be a threshold area for the crack rate below the knee but it shows some scatter. The threshold ΔK exists for the crack rate above the knee and non-propagating cracks are considered to be induced at ΔK lower than the ΔK_{th}, as is inferred from Fig. 5.12.

It is seen that the fretting fatigue strength and the tangential force of the carbon steel are the same in room air of relative humidity of 50–80 % and in dry air of humidity of 4–10 ppm, i.e. ductile steels are not affected by water vapour in fretting fatigue. The crack propagation rate of the carbon steel was therefore determined in argon in order to investigate the effect of oxygen. The crack rate is retarded remarkably by the removal of oxygen compared to that in air while the crack initiation is almost the same as is shown in Fig. 5.14. This may be due to the adsorption effect of oxygen at the crack tip. As a result, the fretting fatigue strength of carbon steel in argon is higher than that in room air or in dry air, but a contributory effect is the

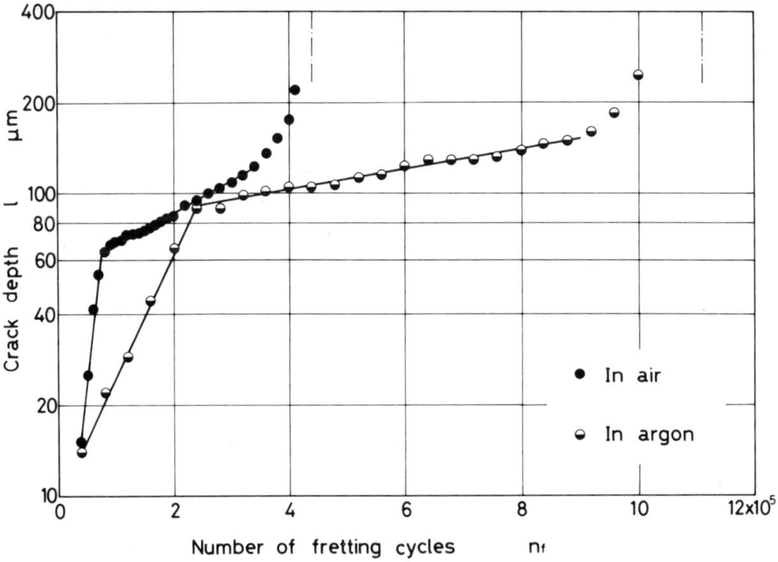

FIG. 5.14. Fretting fatigue crack propagation curves of 0·34 % C carbon steel in air and in argon ($\sigma_0 = 190$ MN m^{-2}, $p = 39$ kN m^{-1}, $f = 30$ Hz).

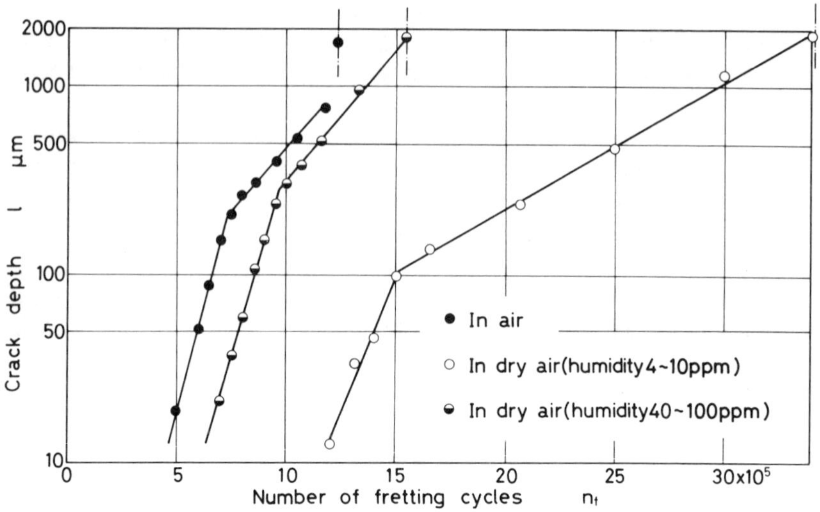

FIG. 5.15. Fretting fatigue crack propagation curves of Al–4Zn–2Mg alloy in air of various humidities ($\sigma_0 = 140\ \mathrm{MN\,m^{-2}}$, $p = 8\ \mathrm{kN\,m^{-1}}$, $f = 20$ Hz).

mechanical factor represented by the variation of tangential force which is known to dominate the fretting fatigue behaviour.

As the aluminium alloy is affected by humidity but is very little affected by oxygen, the crack propagation in dry air was measured. The results are shown in Fig. 5.15. Not only the crack propagation but also the crack initiation are accelerated considerably by a small amount of humidity. The environmental effect predominates over the mechanical effect for the aluminium alloy. In addition, material softening and structural changes of the surface layer were observed.[1] These chemical conditions are also important in the fretting fatigue of aluminium alloys.

5.3 CONCLUSION

Initiation and propagation of fretting fatigue cracks were observed for an annealed carbon steel and an aluminium alloy. Fretting fatigue cracks are initiated at a very early stage of the life. The cracks propagate obliquely to the surface due to the tangential fretting stress combined with the applied repeated stress. When the crack depth reaches a given value which depends on the applied stress and on the contact pressure and frequency of fretting, the crack propagates perpendicularly to the applied repeated stress as the effect of the tangential stress on the contact surface is eliminated. This is well understood from the diagram of crack rate dl/dn_f v. stress intensity factor range ΔK calculated solely from the repeated applied stress. Effects of oxygen and humidity on initiation and propagation of fretting fatigue cracks were evident in both the carbon steel and aluminium alloy.

REFERENCES

1. Waterhouse, R. B. and Taylor, D. E. The initiation of fatigue cracks in a 0·7% carbon steel by fretting, *Wear*, **17**, 139 (1971).
2. Endo, K. and Goto, H. Initiation and propagation of fretting fatigue cracks, *Wear*, **38**, 311 (1976).
3. Endo, K. and Goto, H. Effects of environment on fretting fatigue, *Wear*, **48**, 347 (1978).
4. Wharton, M. H., Taylor, D. E. and Waterhouse, R. B. Metallurgical factors in the fretting fatigue behaviour of 70/30 brass and 0·7% carbon steel, *Wear*, **23**, 251 (1973).
5. Forsyth, P. J. E. Fatigue damage and crack growth in aluminium alloys, *Acta Metallurgica*, **11**, 703 (1963).
6. Nishioka, K. and Hirakawa, K. Fundamental investigations of fretting fatigue (Part 6, Effects of contact pressure and hardness of materials), *Bulletin of the JSME*, **15**, 135 (1972).
7. Endo, K., Komai, K. and Suzuki, Y. Influences of stress cycle frequency on propagation of corrosion fatigue cracks, *Bulletin of the JSME*, **18**, 9 (1975).
8. Endo, K. and Komai, K. Effects of stress wave forms on low cycle corrosion fatigue strength. *Memoirs of the Faculty of Engng., Kyoto University*, **27**, 415 (1965).
9. Hoeppner, D. W. Comments on 'Initiation and propagation of fretting fatigue cracks', *Wear*, **43**, 267 (1977).

6

Environmental Effects in Fretting Fatigue

DAVID W. HOEPPNER

Department of Mechanical Engineering, University of Toronto, Canada

6.1 INTRODUCTION

In the earlier sections of this book it has been indicated that fatigue is a failure mechanism resulting from the application of cyclic loads (or deflections). The fatigue failures almost always nucleate at some physical (and thus mathematical) discontinuity. These discontinuities have been identified as the following:[1]

1. Configuration or geometric discontinuities such as notches, change of section, etc.
2. Inherent material discontinuities such as dislocations, particles, pores, etc.
3. Inherent manufacturing discontinuities such as weld lack of fusion, forging splits, seams, blowholes, etc.
4. Service and maintenance (use) induced discontinuities.

The latter discontinuities, those generated from use, are listed below:

(a) Corrosion
(b) Corrosion fatigue
(c) Stress corrosion
(d) Hot corrosion
(e) Wear
(f) Fretting
(g) Galling
(h) Overloads
(i) Creep
(j) Embrittlement
 Gaseous
 Neutron
 Thermal
 Strain ageing
 Temper
 Ageing
(k) Damage—strikes
(l) Misuse
(m) Inspection damage
(n) Maintenance damage
(o) Repair damage

This book concerns item (f) above, i.e. fretting fatigue. It has been noted in other parts of this book that chemical environment and thermal environment play a role in fretting fatigue. This chapter specifically deals with a discussion of environmental effects on fretting fatigue. References 2–5 are suggested as material that will supplement some of the material presented in the following sections.

6.2 ENVIRONMENTAL EFFECTS IN FRETTING FATIGUE

In reference 5*a* a systems method of viewing corrosion fatigue is presented. This method is suggested to permit an assessment of the numerous factors involved in corrosion fatigue. The challenge of fretting fatigue is considerably more complex than corrosion fatigue because of the contacting surfaces. Figure 6.1 presents a modified systems view of the factors to be considered in assessing both the mechanistic aspects of fretting fatigue and the engineering structural design considerations of fretting fatigue.

As far as the contact stress aspect of fretting is concerned it is generally known[2,4] that the following factors play a role in fretting fatigue:

1. Slip amplitude;
2. Contact stress distribution;
3. Contact materials;
4. Friction coefficient(s);
5. Contact pressure;
6. Surface condition;
7. Environment.

The literature reveals that very little study of the specific role of environment in fretting fatigue has been undertaken. (A later section of this chapter will discuss the studies carried out to date.) Undoubtedly this is due to the extreme complexity of the fretting fatigue experiments themselves, apart from the difficulty of attempting to control or modify either the chemical or thermal environment (or both) during an experimental programme related to fretting fatigue. The fatigue research community knows so little about the phenomenology, mechanisms, and kinetics of corrosion fatigue. Add the further complexity of fretting and a great challenge results.

It is necessary to emphasise the empirical nature of fretting fatigue. Thus, a designer or materials application engineer, faced with a fretting fatigue

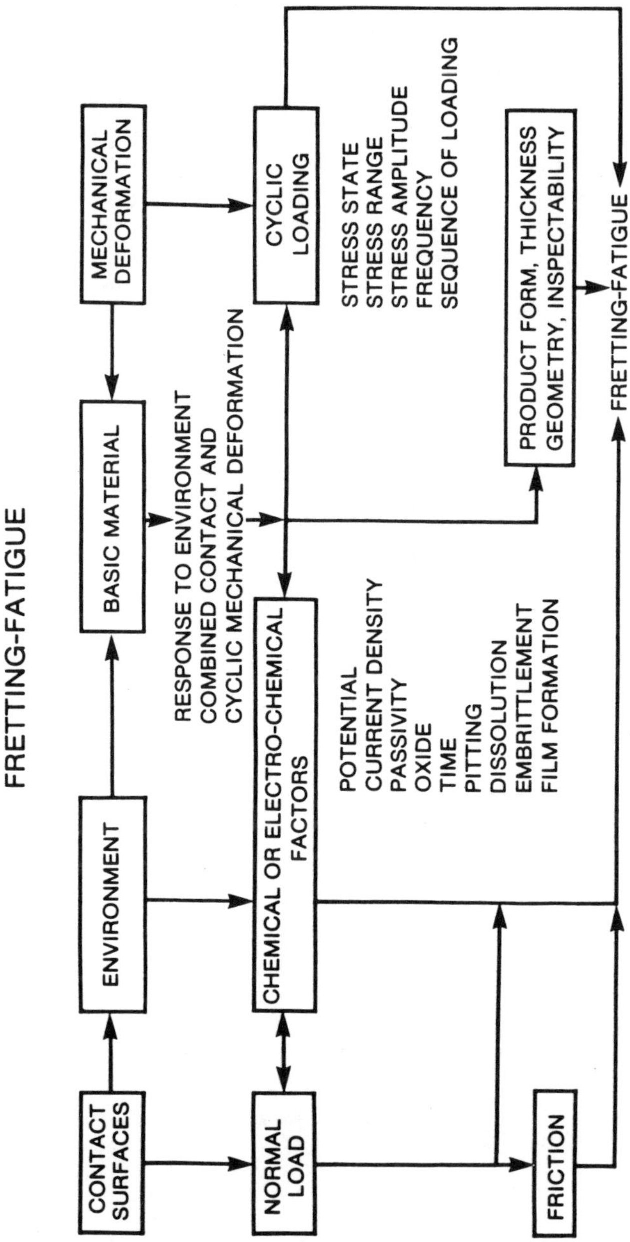

FIG. 6.1. Systems diagram illustrating numerous factors in fretting fatigue.

design problem, is strongly recommended to test the component(s) in question in the anticipated environments. This may not appear desirable at times because of the extensive costs that may be involved; furthermore the 'pure scientist' may wish to predict *a priori* how long a structure subjected to fretting fatigue will last. While the latter aspect is desired it is not possible at the present. Many of the authors of this book have devoted their energy to uncovering the mysteries of fretting fatigue in order to aid the prediction of life of a structure. It should be clear to all readers that we have 'only just begun' to understand fretting fatigue. Using even our limited understanding it is possible to formulate more realistically designs, test programmes, and schemes for alleviation of fretting.[4] Thus, in this section I will attempt to present some guidelines for dealing with the problem of the effect of environment on fretting fatigue.

6.3 THE IMPORTANCE OF THE SURFACE, CONTACT AREA, AND NEAR SUBSURFACE IN FRETTING FATIGUE

In Figs. 6.2 and 6.3 the problem of fretting fatigue is reiterated in conceptual form. Two (or more) surfaces in contact but in 'small' relative motion with respect to one another result in 'damage' to the surface in the contact (slip) region. In addition, damage has been observed to occur in the subsurface region—especially in large grain size materials that can cleave in either an intercrystalline or transcrystalline mode. Figures 6.4 to 6.6 present views of fretting-initiated cracks in one material. From numerous experimental studies conducted under various contact conditions it has been determined that the slip amplitude plays a role in the amount of 'surface damage'. Undoubtedly the slip amplitude is related to friction

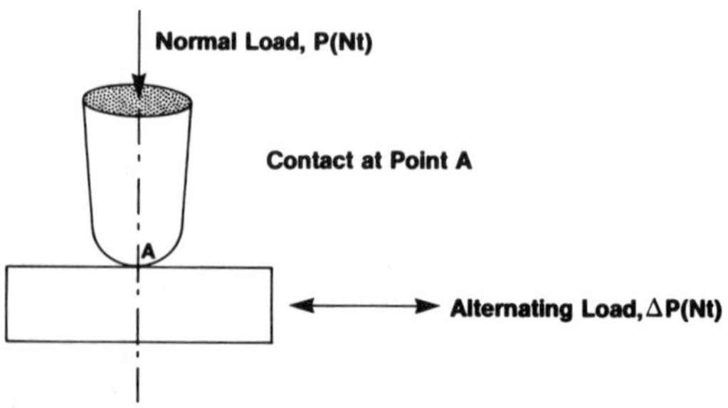

FIG. 6.2. Potential fretting situation—Fretting occurs at contact point A in some environments.

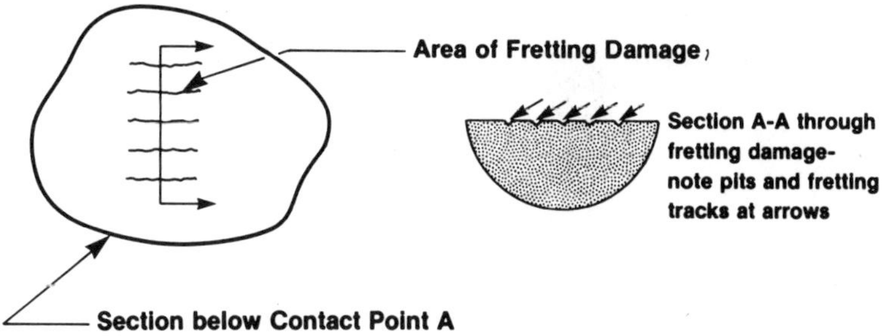

FIG. 6.3. Potential fretting situation—Fretting damage occurs at contact point A. Damage can take form of wear tracks or pits as well as other surface damage types.

coefficient, material, surface finish, surface condition and contact pressure but extensive emphasis has been (and is being) placed upon slip amplitude as a major correlative parameter in fretting fatigue.[2]

From the extensive research conducted on fretting fatigue in the past 15 years, it is now known that a fretting fatigue damage threshold exists[2,6,7] and that after this threshold is passed in the fatigue loading situation the contact has little, if any, effect on the subsequent life. Albeit, the environment may have a significant influence on the fatigue-crack propagation life[5] but the contact condition (fretting) ceases to have an influence after the threshold is crossed. Both the works of Endo,[7] Hoeppner[6] and more recently Alic[8] indicate that after a mode I fatigue crack initiates from the fretting damage the fretting plays no further role in propagating the fatigue crack.

The net result of fretting fatigue is to introduce a reduction in fatigue life at a given stress and also to eliminate the fatigue limit or apparent fatigue limit. These effects are illustrated diagrammatically in Fig. 6.7. The baseline fatigue condition is illustrated (upper band) as are two bands for fretting condition A and fretting condition B. As may be noted condition B is a much more deleterious condition than A. In a later section conditions A and B will be used to illustrate the environmental component in fretting fatigue. The consequence of both of these occurrences is to introduce premature failures or to cause excessive repair/maintenance costs. In the next section some of the theories of fretting are reviewed and discussed in relation to environmental effects in fretting fatigue.

6.3.1 Theories of Fretting

In the many years that fretting has been under study several theories of

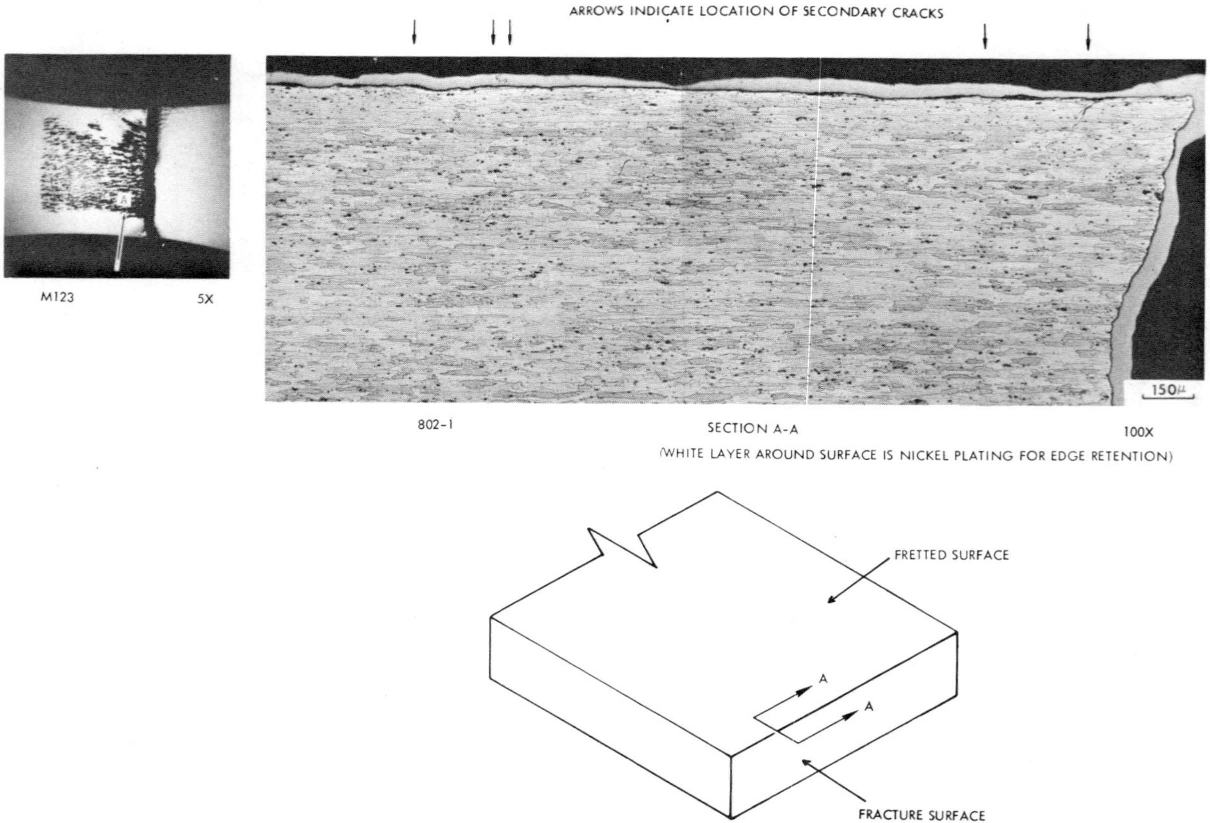

FIG. 6.4. Fretting cracking initiating from surface distress—note multiple crack nucleation sites. Specimen number 71-5A; Material 7075-T6 Aluminium; Maximum fatigue stress—60 000 psi; Normal pressure—3000 psi; Area location—A–A; Distance from specimen edge—0·060 in; Distance from fracture surface—on fracture surface A–A; Distance from fretted surface—on surface. From research of Hoeppner and Goss.[6,9]

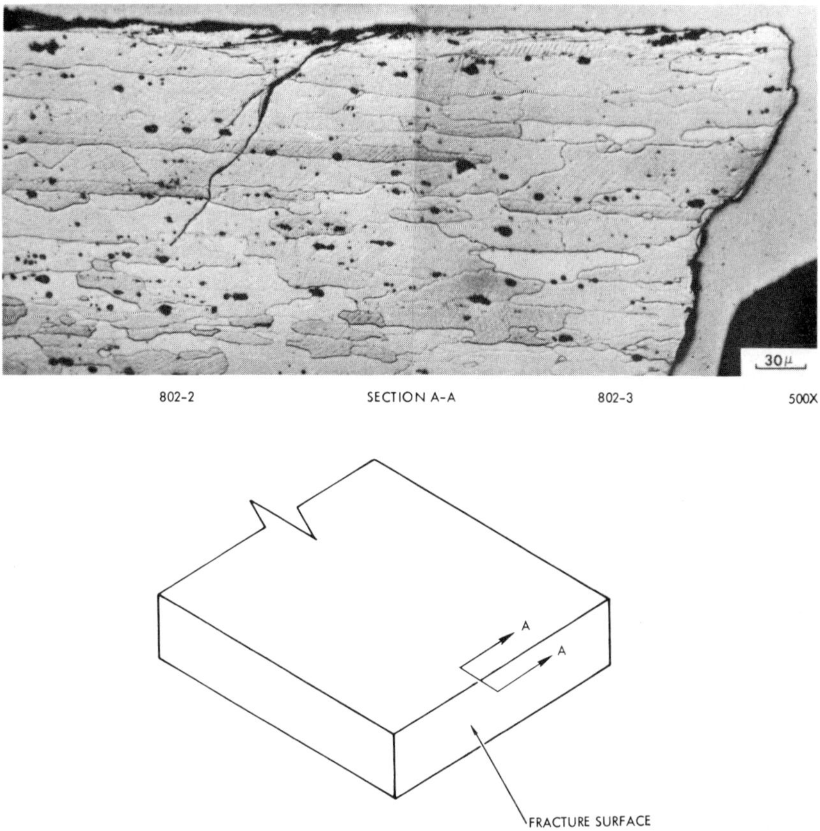

FIG. 6.5. Detail of fretting initiated crack in relation to surface damage of Fig. 6.4. Specimen number 71-5A; Material 7075-T6 Aluminium; Maximum stress—60 000 psi; Normal pressure—3000 psi; Area location—A–A; Distance from specimen edge—0·060 in; Distance from fracture surface—fracture surface–0·010 in; Distance from fretted surface—on surface. From research of Hoeppner and Goss.[6,9]

fretting have been proposed. Several that are judged significant by the author are as follows (see reference 9 for a more detailed discussion):

Theory of Tomlinson[10]

Tomlinson, the first to conduct a systematic study of fretting, hypothesised that fretting resulted from 'molecular attrition'. He further argued that the cohesive forces between surface molecules resulted in tearing away and subsequent oxidation.

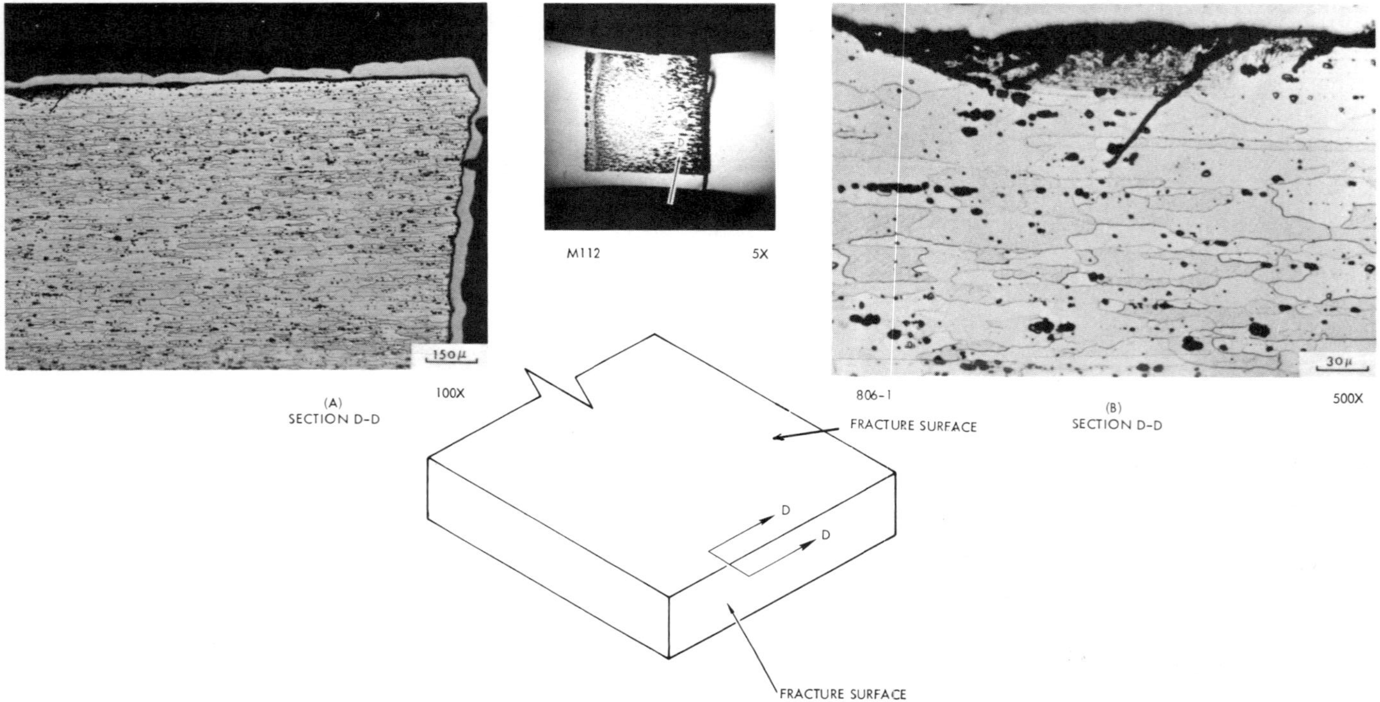

FIG. 6.6. As Fig. 6.4 at lower applied cycle stress. Specimen number 71-1A; Material 7075-T6; Maximum fatigue stress—40 000 psi; Normal pressure—3000 psi; Area location—D; Distance from specimen edge—0·088 in; Distance from fractured surface—0·045 in; Distance from fretted surface—on surface.

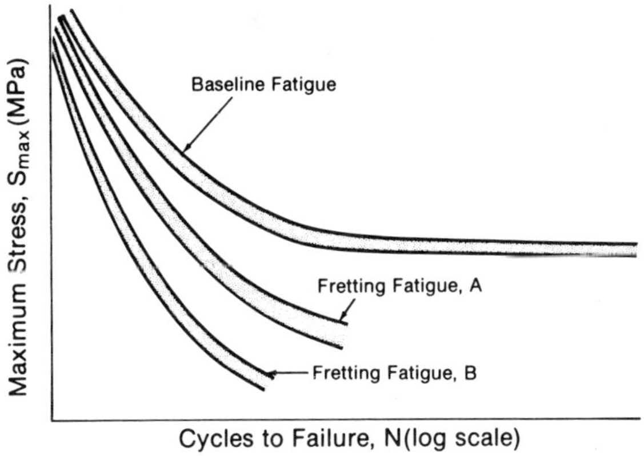

FIG. 6.7. Schematic illustrating the potential reduction in fatigue life brought about by two different test conditions (A and B). Condition B illustrates a much more deleterious effect than condition A.

Theory of Godfrey et al.[11-15]

Godfrey and coworkers employed optical microscopy in attempting to characterise the nature of fretting and establish its mechanism. They concluded that adhesion resulted from contact, and extremely fine particles were subsequently broken loose and oxidised. Another of their findings was that fretting occurred in less than one cycle, consequently they concluded alternating motion was not a necessary condition for fretting and surface fatigue was not a factor.

Theory of Feng and Rightmire[16-19]

Feng. and Rightmire hypothesised a multi-stage fretting process. It consisted of four stages:

(a) Initial stage
(b) Transition period
(c) Declining stage
(d) Steady-state stage

The process they visualised consisted basically of the break-away of particles, subsequent oxidation and abrasion. This process, they believed, reached a steady-state (saturation) level when the oxide layer became very thick.

Theory of Halliday and Hirst[20]

These authors were the first to employ optical and electron microscopy and electron diffraction to the fretting process. They attempted to define the relationship between fretting and magnitude of slip amplitude. Welding of the surfaces occurs, according to their observations, and the degree of relative slip influences the result. They showed that fretting was not dependent on the formation of an oxide.

Theory of Uhlig[21-22]
Uhlig postulated that both a chemical factor and a mechanical factor are involved in fretting. He further proposed a quantitative expression for fretting damage, viz.,

$$W(\text{total}) = (k_0 L^{1/2} - k_1 L)\frac{C}{f} + k_2 lLC \tag{6.1}$$

where W is the specimen weight-loss, L the load, C the number of cycles, f the frequency, l the slip and k_0, k_1, and k_2 are constants. This equation predicts that the fretting weight-loss is a hyperbolic function of frequency.

Theory of Waterhouse[23]
Waterhouse, building on concepts expressed earlier by others, believed fretting was much like wear in that adhesion, the breaking of welds, and metal transfer occurs. In special cases oxides play an important role in the process.

Theory of Wright[24,25]
Wright, principally through electron microscopic observations, emphasised that oxygen was a principal factor causing fretting damage.

It should be noted that several theories presented above attach a great deal of significance to the role of environment in the fretting process. However, very few studies in which environment was the control variable in fretting have been reported. (NASA Lewis Research Center was conducting research in this area but no results have been noted as yet.)

6.3.2 Theories of Fretting Fatigue

No single 'theory' or physical/chemical model of fretting fatigue has emerged that covers all fretting fatigue potential occurrences. Numerous theories of fretting are discussed in reference 4. Almost all the theories or models of fretting concentrate on the role of the fretting in the nucleation of a bonafide fatigue crack from the fretting damage (surface distress) area. The fretting is viewed as providing a local site that causes elevation of strain that ultimately causes *crack* nucleation. The fretting damage is viewed as accelerating the nucleation and early growth of fatigue cracks. Both adhesion and abrasion have been suggested as contributing to the fretting fatigue process. The details of the fretting fatigue process *per se* are discussed in other chapters of this book; additional information can be obtained in references 6–9, and references 25–34 relate the numerous parameters involved in fretting fatigue. When the extensive literature on fretting fatigue is examined it is noted that even though several theories (above) on fretting stress the importance of environment very few studies on the role of environment in fretting fatigue have been conducted.

6.4 THE ROLE OF ENVIRONMENT IN FRETTING FATIGUE

In the other chapters of this book it has been demonstrated how cracks emanate from fretted areas on a surface subjected to cyclic load. It should be obvious that environment must be involved in fretting fatigue since it is involved in fretting (see above) and it is involved in fatigue.[5] There is every reason, therefore, to expect that both chemical and thermal environment must be involved in fretting fatigue. However, as will be seen, no generalisations related to the role of environment can be made at this time.

Endo *et al.*[32] conducted one of the first studies where the role of environment was in part demonstrated. The work cited was mainly concerned with the effect of frequency on fretting fatigue of carbon steel. They showed that decreasing the test frequency decreased fatigue life for a given fretting condition. This result is the same trend observed in environmentally influenced fatigue phenomena. While not a great deal can be concluded from this study related to the specific influence of environment, these results indicated that a chemical (rate controlled) aspect was very important to fretting fatigue. In the last 10 years both Waterhouse *et al.*,[35–39] Hoeppner *et al.*[40,41] and Endo and Goto,[42] have done extensive studies on the role of environment in fretting fatigue. Extensive surveys of this topic can be found in references 43 and 44.

One question that has been of concern in relation to fretting fatigue until recently was related to the potential for fretting fatigue in vacuum. Recent studies[40,41] show that fretting fatigue in vacuum does occur for the vacuum studied. However, fretting fatigue occurs more readily for steel in vacuum than for aluminium. From these works, it is clear that chemical environment plays a major role in fretting fatigue of some materials. Waterhouse *et al.*, Endo *et al.* and Hoeppner *et al.*, have also conducted fretting fatigue in deleterious environments such as salt water and distilled water. For some materials, particularly titanium alloys,[39] the nature of the corrosion product appears to play a major role in the fretting fatigue mechanism. Other materials, such as low carbon steels, do not appear to exhibit as strong a dependence on corrosion product.

The studies performed to date (most completed in the past 10 years) on the role of environment in fretting fatigue have not revealed the mechanism of environmentally affected fretting fatigue. What the studies have revealed is that the chemical environment does play a role in fretting fatigue of some materials for certain test conditions. The results on this aspect are so significant[35–45] from both a mechanistic and structural durability and safety perspective that much greater emphasis will be placed on them in the future.

Returning now to Fig. 6.7 the influence of environment on fretting fatigue can be demonstrated as follows:
Consider the following three cases:

Baseline fatigue—environment A,
Fretting fatigue—environment A,
Fretting fatigue—environment B.

The bands of data indicated in Fig. 6.7 by the shaded area between two solid lines for the three conditions cited above suggest potential effects of fretting including that in a more deleterious environment (environment B). If environment B introduces a more deleterious effect than A then a further reduction of fatigue life is evident. This has been observed for various environments on titanium alloys and for aluminium alloys in air (B) and vacuum (A). From studies on the chemical environment's role in the fretting process it is clear that for many materials, environment may introduce a more deleterious effect. Further studies of this important phenomenon are needed.

Figure 6.8[45] summarises in chart form the many factors to be considered in fretting fatigue. The diagram is intended to be read from left to right—representing the actual physical/chemical process. The materials, stress, environment chemistry and temperature, frequency, wave form, and contact stress must all be considered in both the mechanism of fretting fatigue and structural behaviour for fretting fatigue resistance. As the process occurs (move to right on chart) sufficient 'fretting damage' occurs to provide a localised strain condition that results in development of a mode I, mode II or mode III crack. The crack initiated in the fretting damaged area may propagate. The environment may also influence the subsequent propagation.[5] If the load is continually applied (reversed) until the fretting-initiated crack propagates to a length that is

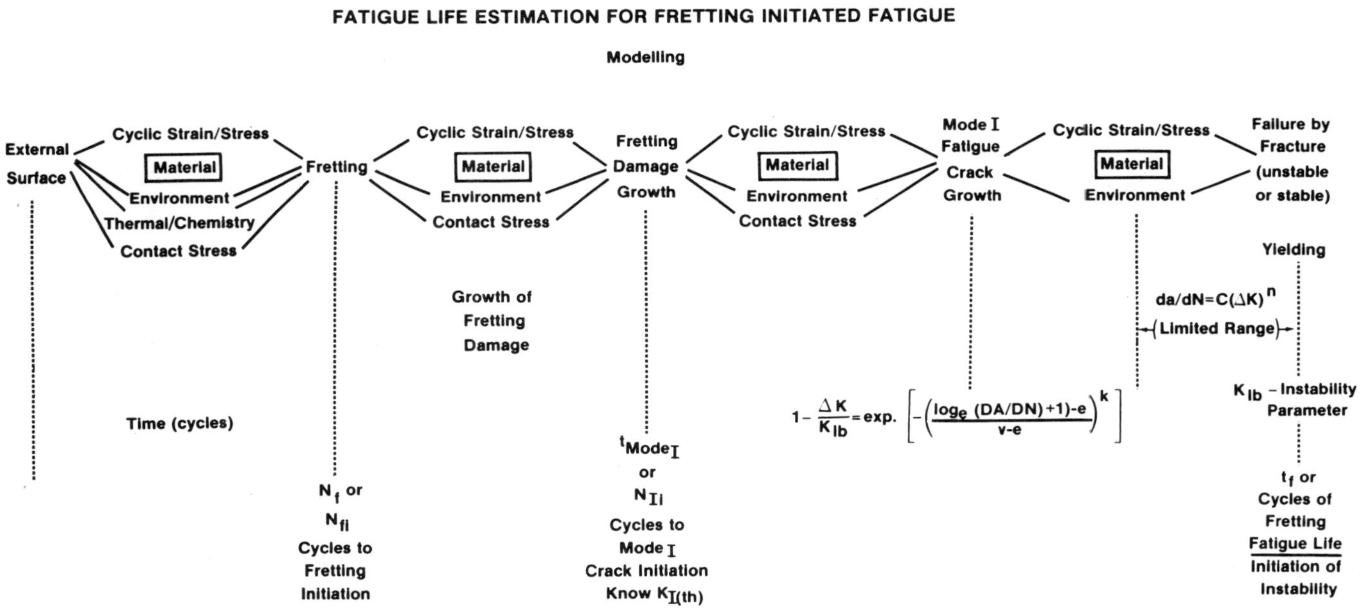

FIG. 6.8. Conceptual chart for fatigue life estimation for fretting initiated fatigue (from reference 45 with permission).

sufficient at the operative stress range, fracture will occur by instability for mode I cracking—right side of diagram. The chart shown in Fig. 6.8 indicates those factors that must be considered in fretting fatigue. Many areas in this chart need much further research so that additional understanding and design guidelines can result.

The role of environment in fretting fatigue is becoming clearer. However, since the process is not currently understood it is suggested that greater emphasis be given to research in this area. Inasmuch as a clear understanding is not available the only recourse to the structural designer and materials application engineer is to test the materials combinations in the structure under conditions that most closely approximate to operation of the structure. Some industries may have extensive information on the role of environment in fretting fatigue and thus know how to transfer data from simplistic fretting fatigue tests under accelerated laboratory tests. Since very little information exists on this subject to date the only reasonable conclusion is that the need remains for a great deal of research in this area. Clearly, if we all wish to reduce the risk of failure in our structures we must get on with the work needed.

REFERENCES

1. Hoeppner, D. W., *Estimation of component life by application of fatigue crack growth threshold knowledge*, Prepared for the Metals Properties Council, New York (In press).
2. Waterhouse, R. B., *Fretting corrosion*, Pergamon Press, Oxford, England (1972).
3. *Specialists meeting on fretting in aircraft systems*, AGARD Conference Proceedings No. 161, NATO-AGARD (1975).
4. *Control of fretting fatigue*, NMAB-333, National Materials Advisory Board, National Research Council, National Academy of Sciences, Washington DC (1977).
5. Devereux, O., McEvily, A. J. and Staehle, R. W. (Eds) *Corrosion fatigue: Chemistry, Mechanics, and Microstructure*, NACE-2, National Association of Corrosion Engineers, Houston, Texas (1972) (*a*) Hoeppner, D. W., Corrosion fatigue considerations in materials selection and engineering design, pp. 3–12, and (*b*) Waterhouse, R. B., The effect of fretting corrosion in fatigue crack initiation, pp. 608–17.
6. Hoeppner, D. W. and Goss, G. L., A fretting-fatigue damage threshold concept, *Wear*, **27**, 61–70 (1944).
7. Endo, K. and Goto, H., Initiation and propagation of fretting fatigue cracks, *Wear*, **38**, 311–24 (1976). *See also* discussion of above with reply: Hoeppner, D. W., Comments on initiation and propagation of fretting fatigue cracks, *Wear*, **43**, 267–70 (1977).

8. Alic, J. A. and Kantimathi, A., Fretting fatigue, with reference to aircraft structures. SAE paper No. 790612, presented at Business Aircraft Meeting and Exposition, Century II, Wichita Ks., April 3–6, 1979, SAE, Wanendale, Pa. 15096.
9. Hoeppner, D. W. and Goss, G. L., Research on mechanism of fretting fatigue, In *Corrosion fatigue* (Devereux, O., McEvily, A. J. and Staehle, R. W., eds) NACE-2, National Association of Corrosion Engineers, Houston, Texas, pp. 617–26 (1972).
10. Tomlinson, G. A., The rusting of steel surfaces in contact, *Proc. Roy. Soc.*, **115A**, 472 (1927).
11. Godfrey, D., Investigation of fretting corrosion by microscopic observation, NACA Technical Note 2039, February 1950.
12. Godfrey, D., A study of fretting wear in mineral oil, *Lubrication Engineering*, **12**, January–February, 37–42 (1956).
13. Godfrey, D. and Bailey, J. M., Coefficient of friction and damage to contact area during the early stages of fretting—I. Glass, copper or steel against copper, NACA Technical Note 3011, September 1953.
14. Godfrey, D. and Bailey, J. M., Early stages of fretting of copper, iron, and steel, *Lubrication Engineering*, **10**, May–June, 155–9 (1954).
15. Godfrey, D. and Bisson, E. E., NACA studies of mechanism of fretting (fretting corrosion) and principles of mitigation, *Lubrication Engineering*, **8**, October, 241–3 (1952).
16. Feng, I-Ming, Fundamental study of the mechanism of fretting, MIT Lubrication Lab, Cambridge, Mass., Final Report AD59089, February 28, 1955.
17. Feng, I-Ming and Rightmire, B. G., *The mechanism of fretting*, Mass. Inst. of Tech., Cambridge, Mass., AD 4463 (1952).
18. Feng, I-Ming and Rightmire, B. G., The mechanism of fretting, *Lubrication Engineering*, **9**, 134–6, 158–61 (1953).
19. Feng, I-Ming and Rightmire, B. G., 'An experimental study of fretting', *Proc. Inst. Mech. Engrs.*, London, **170**, 1055–60 (1956).
20. Halliday, J. S. and Hirst, W., The fretting corrosion of mild steel, *Proc. Roy. Soc.*, London, **236A**, 411–25, August 2, 1956.
21. Uhlig, H. H., Mechanism of fretting corrosion, *Journal Applied Mechanics*, **21**, December, 401–7 (1954).
22. Uhlig, H. H. Feng, I-Ming, *et al.*, Fundamental investigation of fretting corrosion, NACA Technical Note 3029 (1953).
23. Waterhouse, R. B., Fretting corrosion, *Proc. Inst. Mech. Engrs.*, **169**, 1159–72 (1955).
24. Wright, K. H. R., An investigation of fretting corrosion, *Proc. Inst. Mech. Engrs.*, **1B**, 556–71 (1952–1953).
25. Wright, K. H. R., Discussion and communication on an investigation of fretting corrosion, *Proc. Inst. Mech. Engrs.*, **1B**, 571–4 (1952–1953).
26. Nishioka, K., Nishimura, S. and Hirakawa, K., Fundamental investigations of fretting fatigue—Part 1, On the relative slip amplitude of press-fitted axle assemblies, *Bulletin of JSME*, **11**, No. 45, 437–45 (1968).
27. Nishioka, K. and Hirakawa, K., Fundamental investigation of fretting fatigue —Part 2, Fretting fatigue testing machine and some test results, *Bulletin of JSME*, **12**, No. 50, 180–7 (1969).

28. Nishioka, K. and Hirakawa, K., Fundamental investigations of fretting fatigue —Part 3, Some phenomena and mechanisms of surface cracks, *Bulletin of JSME*, **12**, No. 51, 397–407 (1969).
29. Nishioka, K. and Hirakawa, K., Fundamental investigations of fretting fatigue —Part 4, The effect of mean stress, *Bulletin of JSME*, **12**, No. 51, 408–14 (1969).
30. Nishioka, I. and Hirakawa, K., Fundamental investigations of fretting fatigue —Part 5, The effect of relative slip amplitude, *Bulletin of JSME*, **12**, No. 52, 692–7 (1969).
31. Nishioka, K. and Hirakawa, K., Fundamental investigations of fretting fatigue —Part 6, Effects of contact pressure and hardness of materials, *Bulletin of JSME*, **15**, No. 80, 135–44 (1972).
32. Endo, K., Goto, H. and Nakamura, T., Effects of cycle frequency on fretting fatigue life of carbon steel, *Bulletin of JSME*, **12**, No. 54, 1300–8 (1969).
33. Endo, K., Goto, H. and Nakamura, T., Fretting fatigue strength of several materials combinations, *Bulletin of JSME*, **16**, No. 92, 143–50 (1973).
34. Endo, K., Goto, H. and Fukunaga, T., Behaviors of frictional force in fretting fatigue, *Bulletin of the JSME*, **17**, No. 108, 647–54 (1974).
35. Waterhouse, R. B. and Taylor, D. E., The relative effects of fretting and corrosion on the fatigue strength of a eutectoid steel, *Wear*, **15**, 449–51 (1970).
36. Waterhouse, R. B., Dutta, M. K. and Swallow, P. J., Fretting fatigue in corrosion environments, *Proceedings of the Conference on the Mechanical Behavior of Solids*, Kyoto, Japan, 1971, pp. 294–300.
37. Waterhouse, R. B. and Dutta, M. K., The fretting fatigue of titanium and some titanium alloys in corrosive environment, *Wear*, **25**, 171–5 (1973).
38. Waterhouse, R. B. and Wharton, M. H., The behavior of three high strength titanium alloys in fretting fatigue in a corrosive environment, *Lubrication Engineering*, **32**, 294 (1976).
39. Wharton, M. H. and Waterhouse, R. B., Environmental effects in fretting fatigue of Ti–6Al–4V, *Wear*, **62**, 287–97 (1980).
40. Reeves, R. K. and Hoeppner, D. W., Microstructural and environmental effects on fretting fatigue, *Wear*, **47**, 221–9 (1978).
41. Poon, C. and Hoeppner, D. W., The effect of environment on the mechanism of fretting fatigue, *Wear*, **52**, 175–91 (1979).
42. Endo, K. and Goto, H., Effect of environment on fretting fatigue, *Wear*, **48**, 347 (1978).
43. Reeves, R. K., Microstructural and environmental effects on fretting fatigue, Ph.D. Dissertation, University of Missouri, Columbia, May 1977.
44. Poon, C. J., Environmental Effects on the Mechanism of Fretting Fatigue in 7075-T6 Aluminium, Ph.D. Dissertation, University of Missouri, Columbia, December 1978.
45. Hoeppner, D. W., *Structural integrity principles applied to engineering design*, Wiley–Interscience (1980, in press).

7

Fretting Fatigue in Aqueous Electrolytes

R. B. WATERHOUSE

Department of Metallurgy and Materials Science, University of Nottingham, UK

In practice there are many examples of machinery or devices operating in aqueous environments with the possibility that fretting may occur at contacting surfaces as a result of cyclic stressing or vibration. The press-fit or shrink-fit of a hub or bearing housing on to a loaded rotating shaft is in itself a frequent source of fatigue cracks initiated by fretting, but the situation is further aggravated if it is operating, as in many marine applications, in sea water or salt spray. A particular example is that of a ship's propeller shrunk on to the tail-shaft. It is reported that during the first 60 years of this century, failures of tail-shafts due to salt-water corrosion-fatigue have been reduced but the incidence of fretting-corrosion fatigue failures has increased.[1] In a study of 2500 shafts in large single-screw ocean-going vessels over a 10 year period, 22% of those removed for inspection had to be renewed. In about half of them the cause was the presence of fatigue cracks, most of which occurred under the propeller hub or under the bronze liner fitted on to the shaft or in the keyway. The evidence is that such cracks develop and propagate under quite moderate loads rather than under the occasionally abnormally high loadings experienced in rough weather. This is a feature of corrosion fatigue and fretting fatigue failures.

A similar application where fretting has been experienced is the mechanical seal on a ship's propeller shaft. There are many designs of such seals but in their simplest form they consist of a seal ring which is fixed to the shaft and spring-loaded against a stationary face which is located on the hull of the ship. Slight misalignment or vibration of the shaft results in axial oscillation of the seal ring which is transmitted to the shaft packing and leads to fretting of the surface of the shaft.[2]

Besides shipping, there are other situations involving a marine environment. At the present time there is considerable interest in generating energy from wave-power. A wide variety of devices have been proposed for achieving this. Some involve relative movement between adjacent portions of hinged floating platforms, others the change in air pressure in moored inverted bells. All will involve the conversion on-site to a more convenient form of energy, e.g. electrical. The devices themselves and the generators will all involve moving parts where fretting may occur. In addition, they will have to be moored and fretting is a possibility between the strands of steel ropes under tension where vibration occurs in flowing water. Also, the electrical energy generated will need to be conducted ashore in flexible cables; these too are multi-component and will also suffer movement and vibration.

A further, quite different, application where fretting has been observed to contribute to fatigue failure in an aqueous environment, is the orthopaedic implant. These may be broadly divided into two groups, (a) the fracture fixation devices such as bone-plates and Kuntscher nails and (b) the partial or total joint replacements. The bone-plate consists of a metal plate, usually austenitic stainless steel, containing a series of screw holes which is attached to the bone via metal screws of the same material as the plate so as to hold the two faces of a fracture together to allow it to heal. Metal-to-metal contact occurs between the bevelled under-side of the screw head and the countersink surrounding the hole. This is obviously a situation where crevice corrosion is likely to occur since the body fluid is broadly an oxygenated 1 % sodium chloride solution. To counter crevice corrosion, the under-side of the screw head is a section of a sphere so that a line contact is made with the countersink. As the limb is exercised, small movements occur at this contact and fatigue cracks are initiated by fretting leading to failure at the hole (Fig. 7.1). Another orthopaedic example seen by the author was a fatigue failure of the femoral stem of a Charnley hip prosthesis. The operation involves replacing the ball-and-socket joint of the hip with a metal ball riding in a high density polyethylene socket (Fig. 7.2). The metal femoral component has a stem some 5–6 in in length which is cemented into the cavity of the femur with the polymethylmethacrylate cement. In carrying out the operation, some surgeons remove the protruding portion of bone known as the greater trochanter and then reattach it with a stainless steel wire which is anchored by winding it around the stem of the prosthesis. In the particular case, movement occurred between the wire and the stem, producing fretting which led to a fatigue failure at this point.[3] Because of this danger this procedure is no longer encouraged.

Another possibility in hip replacements is that loosening may occur in the cement fixation. When this occurs in the upper portion of the stem so that the femoral component is only supported at the lower end, the bending

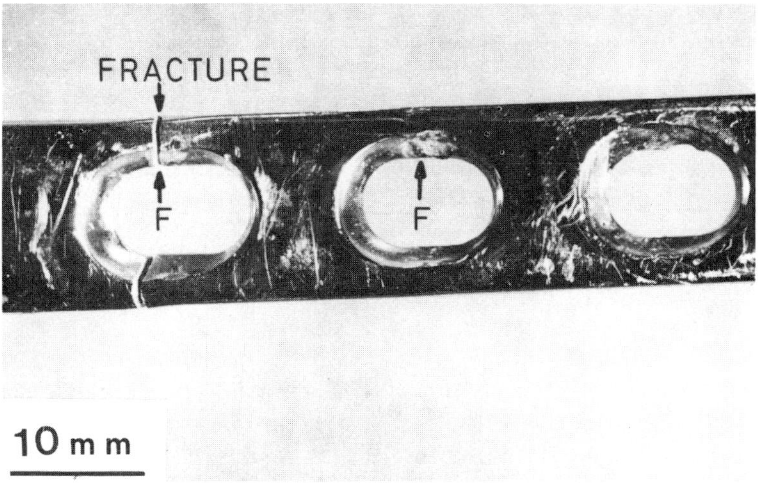

FIG. 7.1. Fatigue fracture of a bone-plate caused by fretting between the underside of the screw head and the countersink. Points marked F indicate fretting damage.

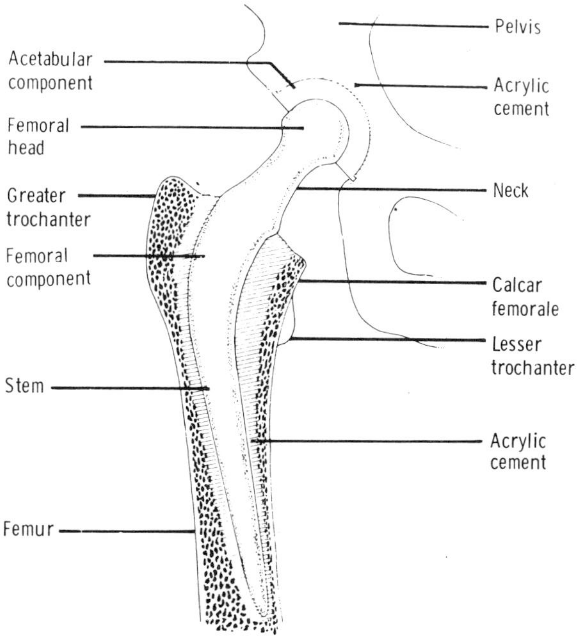

FIG. 7.2. Schematic diagram of artificial hip joint.

moment on the stem is increased with the enhanced probability of fatigue failure and even plastic deformation. A further possible hazard is that fretting may occur between the cement and the stem and contribute to fatigue failure. This type of fretting has been investigated in the author's laboratory and is dealt with below.

These few examples illustrate the diversity of the situations in which fretting in an aqueous environment can be a factor in fatigue failure.

7.1 TRIBOLOGY IN AQUEOUS CONDITIONS

When a clean oxide-free metal surface is immersed in an electrolyte, positively charged metal ions pass into the solution leaving the surface with a net negative charge. The positive ions attract water molecules to themselves forming positively charged aquo-cations which are attracted to the negatively charged surface forming the electrical double layer. Very rapidly the electric field in the double layer leads to a state of dynamic equilibrium where the number of metal ions leaving the surface is balanced by those being redeposited on the surface from the solution. Outside the double layer there is a concentration gradient of the positively charged aquo-cations into the bulk of the solution. The potential of the metal relative to the solution, the so-called electrode potential, depends on the position of the metal in the electrochemical series, the nature and concentration of the electrolyte, and other reactions which may occur at the interface. In practice, in neutral solutions most metals develop a hydroxide or oxide film. The extent to which this grows depends on its ionic and electronic conductivity. In some cases the film is not protective, e.g. as in the rusting of mild steel in sodium chloride solution, while in others the film is extremely thin and highly protective, e.g. the passive film formed on stainless steels and titanium alloys.

The presence of such films and also the physical adsorption of various species has a profound effect on the sliding and coefficient of friction of metal surfaces in contact with each other. This was first demonstrated by measuring the coefficient of friction between two platinum surfaces sliding in 0·1 N sulphuric acid as a function of the electrode potential of the couple.[4] Maximum friction was observed at the isoelectric point when neither hydrogen ions nor oxygen ions were being adsorbed on the surface and the interfacial energy between the platinum and the electrolyte was a maximum. As the potential was lowered, hydrogen ions were adsorbed, and as it was raised oxygen ions were adsorbed. The latter produced the greater reduction in friction.

In the absence of an oxide film and in a vacuum, very high coefficients of friction are recorded since the real area of contact between the surfaces progressively increases as shearing occurs. It is the presence of the oxide film with its very different mechanical properties which results in the typical coefficients of friction for metal-on-metal sliding of around 0·5 to 0·8.

The other aspect of the problem is the effect the sliding action has on the oxide films themselves. On corrosion-resistant materials such as aluminium alloys, titanium alloys and stainless steels, the oxide film is highly protective, but mechanical disruption will momentarily expose the underlying metal to the corrosive environment and the consequent reaction will to a greater or lesser extent repair the damage. Where the process is repeated as in continuous sliding or in the reciprocating motion of fretting, a higher than normal rate of wear may be the result. This would be an example of corrosive wear. It has been found in the textile industry that the wear of stainless steel guides over which yarn was sliding in an oil-in-water emulsion was greater when the potential of the steel surface was maintained in the passive region, so that the oxide film was continually reformed when removed, compared with the cathodic region where no oxide film was present.[5]

For a more detailed treatment of the interaction between tribology and electrochemical behaviour reference should be made to an article by the author.[6]

7.2 THE EFFECT OF ENVIRONMENT ON FRETTING FATIGUE

Fretting is mainly concerned with initiating fatigue cracks whereas corrosion is concerned with both initiation, in attacking emerging slip bands, and propagation, by similar processes at the tip of the growing crack. It is of interest, therefore, to consider the possible interaction between fretting and corrosion at the initiation stage. It is conceivable that the corrosion products are such as to result in a reduction in the coefficient of friction and act in the same way as an artificially applied coating. This has been found in high-temperature oxidation and is dealt with in the following chapter. In aqueous solutions this does not generally appear to be the case, unless a specific film-forming inhibitor is added to the solution. Addition of 10 % potassium chromate to a 0·1 N sodium chloride solution more than doubled the fretting-fatigue life of an Al–6Zn–3Mg alloy.[7] There was further evidence that this concentration of the inhibitor prevented the initiation of a fatigue crack in the first 50 000 cycles of fretting.

Once the oxide film has been disrupted, then some form of conjoint action between the fretting and the corrosive environment may occur. It should be pointed out, however, that the common experience from the type of crack-growth studies described in Chapter 1 is that at higher stress levels and higher crack-growth rates the environment has less effect. Fretting is thought to be a form of local high-strain fatigue and, therefore, it might be thought that the initial growth of the crack would not be influenced by the environment. However, the work of Endo (see Chapter 5) indicates that the initial rate of growth is higher in air than in argon. When the crack propagates beyond the region immediately affected by the fretting then the effect becomes very apparent (see Fig. 5.14).

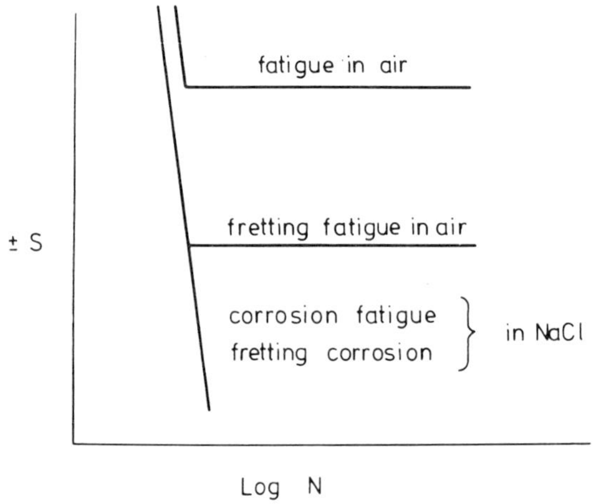

FIG. 7.3. Schematic diagram of S–log N curves for a non-corrosion-resistant material, e.g. mild steel, in air and saline with and without fretting.

In general, materials can be divided into two groups with regard to their response to fretting fatigue in an aqueous electrolyte. This is illustrated diagrammatically in Figs. 7.3 and 7.4 which represent the type of S–log N curves for materials which are not corrosion-resistant in the particular environment (Fig. 7.3) and those that are (Fig. 7.4). In the first group, e.g. mild steel in sodium chloride solution, when fretting and corrosion are both present, the behaviour is entirely dictated by the corrosion properties and fretting produces no further reduction in the fatigue strength. The second group experience little or no reduction in fatigue strength due to corrosion alone but a large reduction when fretting and corrosion are present even

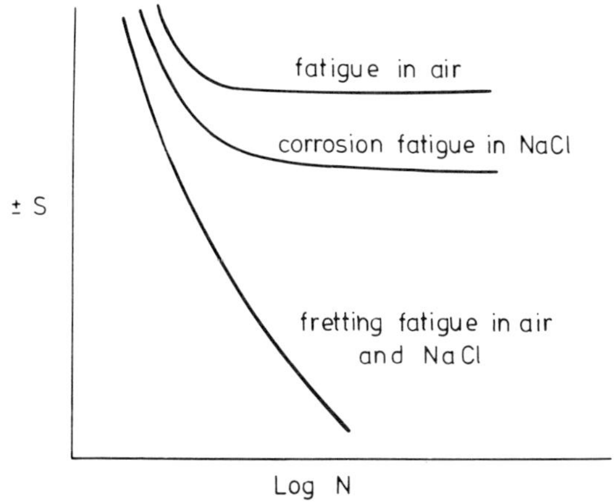

FIG. 7.4. Schematic diagram of *S*–log *N* curves for a corrosion-resistant material, e.g. austenitic stainless steel, in air and saline with and without fretting.

though the corrosive environment is relatively mild, e.g. air. In this case the prime function of the fretting is the disruption of the otherwise protective oxide film.

Measurements have been made of the fall in potential when surfaces are fretted. This fall occurs due to the exposure of the metal surface and the rapid escape of metal ions into the solution. The extent of the potential drop gives some indication of the reactivity of the surface. In a study of the fretting of a number of pure metals in 0·5 M sodium chloride, falls of potential shown in Table 7.1 were obtained.[8] From these results it is

TABLE 7.1
THE MEAN FREE POTENTIAL FALL ON FRETTING METALS IN 0·5 M NaCl

Metal	*Fall in free potential (mV)*
Silver	0
Copper	0
Nickel	35
Tantalum	130
Chromium	310
Zinc	10
Titanium	500
Aluminium	450

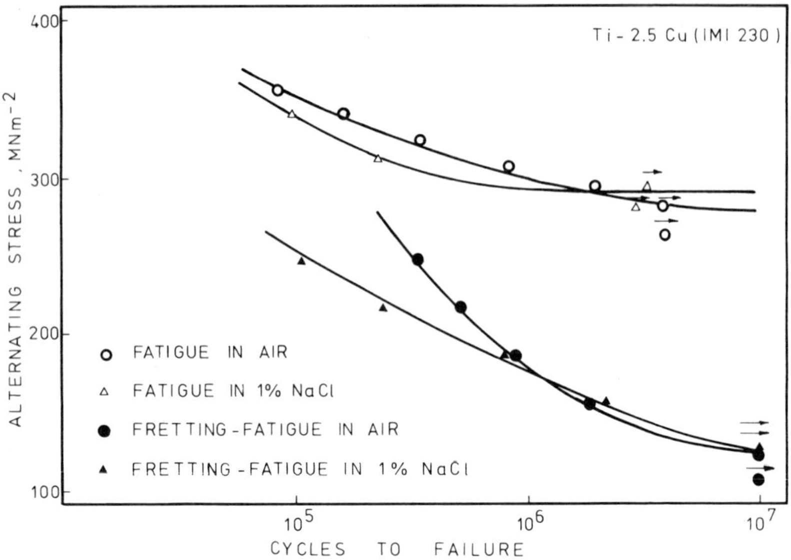

FIG. 7.5. Effect of fretting on the fatigue behaviour of the titanium alloy Ti–2·5Cu tested in rotating bending.

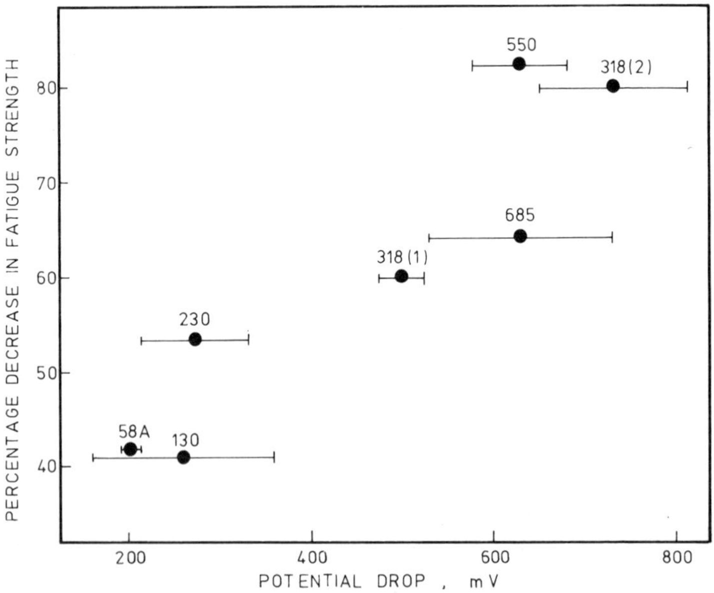

FIG. 7.6. Relation between potential drop and reduction in fatigue strength of a number of titanium alloys when fretted in 1% NaCl. Key: 130 commercial purity Ti; 230, Ti–2·5Cu; 318, Ti–6Al–4V; 550, Ti–4Al–4Mo–2Sn–0·5Si; 685, Ti–6Al–5Zr–0·5Mo–0·25Si, En 58A 18Cr–10Ni austenitic stainless steel.

apparent that the metals tantalum, chromium, aluminium and titanium and their alloys are likely to be most susceptible to fretting–corrosion fatigue. In practice it has been found that there is some correlation, if not an exact one, between the potential drop in such experiments and the reduction in strength in the fatigue test. Figure 7.5 shows the effect of fretting on the fatigue behaviour of a titanium alloy.[9] In common with most titanium alloys this alloy is little affected by corrosion fatigue but fretting reduces the fatigue strength by 55 % whether the environment is air or 1 % sodium chloride solution. The percentage reduction in fatigue strength produced by fretting of a number of titanium alloys in 1 % sodium chloride solution is shown in Fig. 7.6, together with the associated falls in electrode potential. The corresponding value for an austenitic stainless steel is also included. As mentioned above, there appears to be some correlation between the fall in potential and the reduction in fatigue strength.

In an earlier monograph some results were quoted which showed that in potentiostatic experiments, the corrosion current increase due to fretting on an austenitic stainless steel in a 1 % solution of sodium chloride was

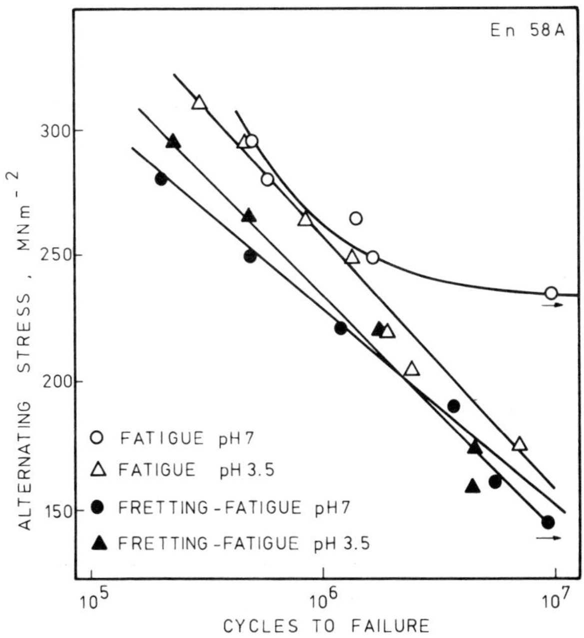

FIG. 7.7. Effect of pH on the fatigue and fretting fatigue behaviour of an austenitic stainless steel in 1 % NaCl.

increased tenfold when the pH was decreased from 7 to 3·5.[10] The effect of pH change has been investigated in the fretting fatigue of this steel.[7] Figure 7.7 shows the fatigue and fretting fatigue curves at the two values of pH. The corrosion fatigue strength shows a substantial reduction at the lower pH but the fretting fatigue behaviour is unaffected. This suggests that disruption of the normally protective oxide film by fretting exposes a very reactive surface to oxygen and water, and the degree of corrosiveness of the environment has little further influence. This would explain why the reduction in fatigue strength due to fretting on corrosion-resistant alloys is about the same in air, which contains water vapour, as in sodium chloride solution.

7.3 EFFECT OF FREQUENCY

It was shown earlier that in potentiostatic experiments on stainless steel in 1% sodium chloride solution the current change produced by fretting was proportional to the frequency.[10] One would therefore expect frequency to have an effect on the potential change described above. This is found to be the case. The potential trace with time is not always constant but can show a gradual fall after the initial large decrease, as in Fig. 7.8. The explanation for this is given in Fig. 7.9 which represents the change in potential as

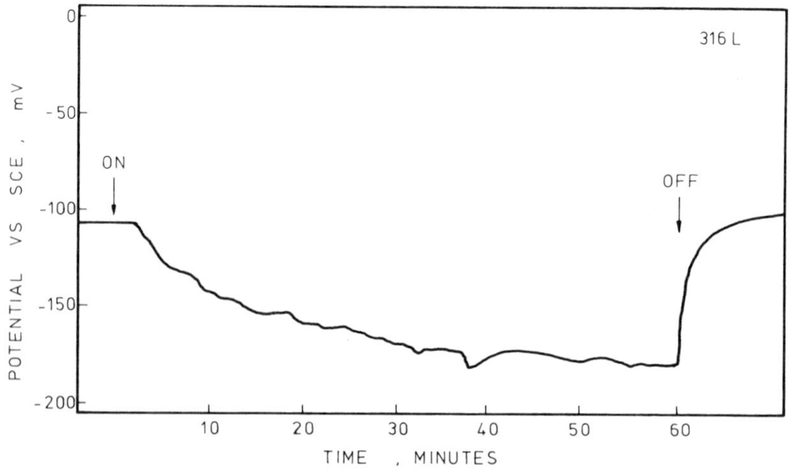

FIG. 7.8. Potential trace on fretting a sphere (diam. 12·5 mm) on flat of austenitic stainless steel 316L in Hanks solution (artificial body fluid) at a frequency of 1 Hz, amplitude of 50 μm and normal load of 0·1 kg.

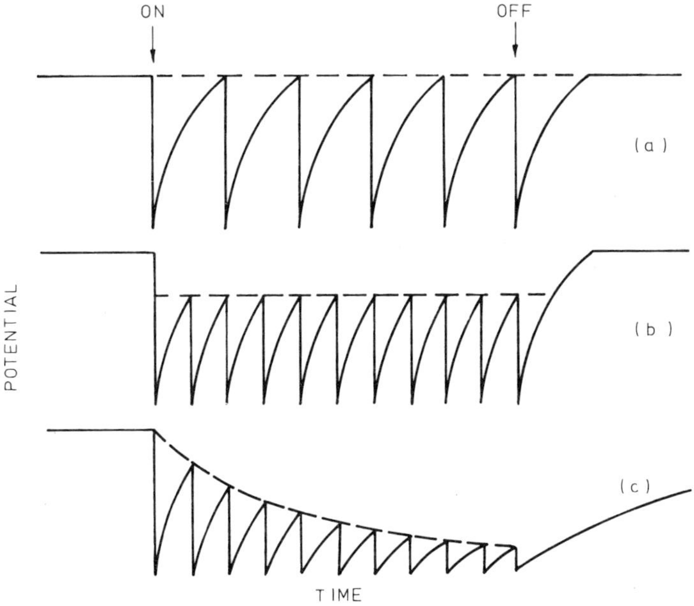

FIG. 7.9. Schematic representation of potential changes due to fretting and repassivation (a) low frequency, (b) high frequency, (c) high frequency with change in rate of repassivation.

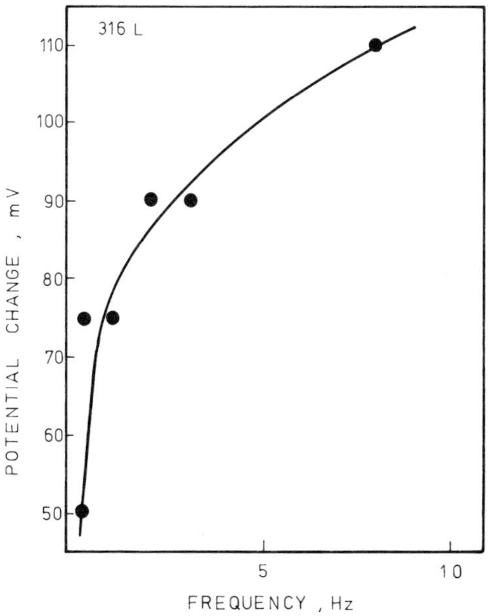

FIG. 7.10. Effect of frequency of fretting on the maximum potential change when an austenitic stainless steel 316L is fretted in Hanks solution—low frequencies.

fretting proceeds. Figure 7·9(a) represents the situation where the damaged surface during one passage of the fretting slider completely repassivates before the slider returns in the next half cycle. In this case little or no fall in potential would be expected. Figure 7.9(b) represents a higher frequency where the surface does not completely repassivate before it is disrupted again. In this case a constant decrease in potential with time is to be expected. In Fig. 7.9(c) not only is repassivation incomplete but the rate of repassivation decreases as the number of fretting cycles increases. This is to be expected in practice since the surface is becoming more and more disrupted. The result in this case is a decrease in potential with time as seen in Fig. 7.8. Figure 7.10 shows the maximum fall in potential in such an experiment plotted against frequency for low frequencies, such as could be expected in body movements. Figure 7.11 relates these potential changes to those at a much higher frequency, namely 150 Hz. The relation between frequency and potential drop is logarithmic.

As yet there has been little investigation of the effect of frequency on fretting fatigue but Fig. 7.12 shows some results obtained on stainless steel

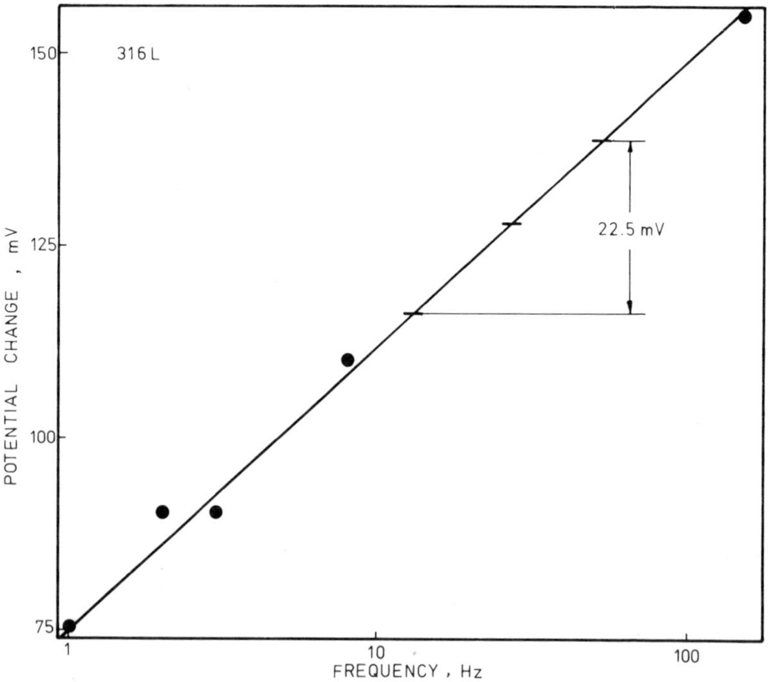

FIG. 7.11. Effect of frequency of fretting on the maximum potential change when an austenitic stainless steel 316L is fretted in Hanks solution—higher frequencies.

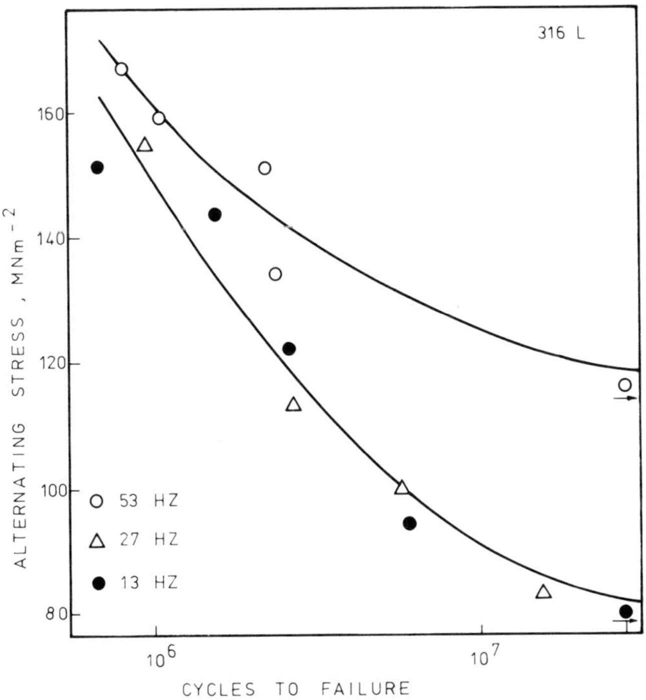

FIG. 7.12. Effect of frequency on the fretting fatigue behaviour of an austenitic stainless steel 316L in Hanks solution tested in push–pull—mean stress 247 MN m^{-2}.

in the author's laboratory. The result appears contrary to the relation between potential drop and fatigue strength shown in Fig. 7.6 since the tests at the higher frequency, where the potential drop would be higher, show a greater fatigue strength. The difference in frequency in these tests would give a maximum difference in potential drop of 22·5 mV as indicated in Fig. 7.11. From Fig. 7.6 this would only result in a very small change in fatigue strength. It is more likely that the differences in Fig. 7.12 are the effect of frequency on corrosion fatigue (see Fig. 1.3) but further work at lower frequencies is required to check this.

7.4 NON-METALS IN CONTACT WITH METALS

In the results described in the foregoing sections the fretting has been between metal specimens and bridges made of the same material. In recent years the effect of various polymers on the fretting wear of steel has been investigated. Damage is caused to the steel through the transfer of oxide films to the polymer and the polymer suffers wear damage due to fibre formation.[11] The damage to the steel surface is much increased in moist conditions (R.H. 85%).[12] No assessment was made of the effect of such materials on the fretting fatigue strength of the steel in air or aqueous environment.

Examination of femoral components from artificial hips removed from patients showed that there were often patches of damage near the upper end of the stem, apparently due to fretting between the metal and the bone cement which had debonded (see Fig. 7.2). Although there were no reported fatigue failures arising from this type of damage it was decided to investigate it by carrying out fretting fatigue tests on a stainless steel and a titanium alloy with bridges of bone cement, which is a polymethylmethacrylate polymer, in contact with the metal fatigue specimens in Hanks solution, which is artificial body fluid consisting essentially of 1% sodium chloride solution with other minor additions. The experiments were carried

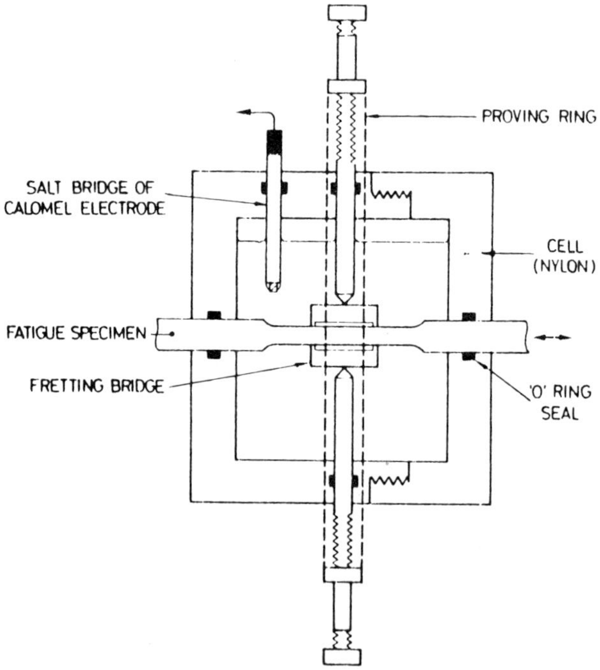

FIG. 7.13. Nylon cell for carrying out fretting fatigue tests in push–pull in an aqueous environment.

out in fluctuating tension in a specially designed nylon cell (Fig. 7.13). The level of stressing was chosen such that if metal bridges had been used, fatigue failure would have occurred within 10^5 to 2×10^5 cycles. Both materials were unbroken after 10^7 cycles of fretting fatigue against the bone cement.[13] It was concluded that there was little effect on fatigue strength. There were changes in potential but these were more transient than in metal-on-metal fretting. The main effect of fretting was to cause thickening and cracking of the oxide film.

7.5 EFFECT OF CATHODIC PROTECTION

The evidence from earlier investigations is that fretting of metal surfaces in aqueous electrolytes increases the corrosion current, although the effect on fatigue strength is very little different whether the tests are carried out in sodium chloride solution or air. One of the effective ways of countering corrosion is cathodic protection.

Some tests were carried out on an austenitic stainless steel in 0·1 M sodium chloride solution in fluctuating tension in the cell described previously. A mean stress of 246 MN m^{-2} and an alternating stress of ± 154 MN m^{-2} were applied. In the tests where cathodic protection was imposed the potential of the specimen was maintained at −1200 mV with respect to the saturated calomel electrode. The results are shown in Table 7.2.[14] These results show that not only does cathodic protection have a significant effect on fatigue life in corrosion fatigue, which was to be expected, but it is also capable of a similar effect where fretting is occurring. This is an important result since it indicates that the mechanical damage produced by fretting on a corrosion-resistant material, such as stainless steel, in a corrosive medium

TABLE 7.2
FATIGUE LIVES OF STEEL En 58A IN AIR AND 0·1 M NaCl, WITH THE EFFECT OF FRETTING AND CATHODIC PROTECTION

Fatigue in air	*Fretting fatigue in air*	*Corrosion fatigue in 0·1 M NaCl*		*Fretting–corrosion fatigue in 0·1 M NaCl*	
		Free potential	*Cathodic protection*	*Free potential*	*Cathodic protection*
$>10^7$	463 000	314 000	2 046 000	286 000	2 279 000

has very little effect on the fatigue strength and therefore the main effect is due to chemical attack. This confirms to some extent the observation by Fenner and Field that a propagating fatigue crack was produced by fretting an aluminium alloy in air after 240 000 cycles but under the same stressing conditions *in vacuo* 2–3 million cycles were necessary.[15] Further recent work of Poon and Hoeppner showed that the fatigue life under fretting conditions of an aluminium alloy *in vacuo* was 10 to 20 times longer than in laboratory air.[16] These authors ascribed the lower fatigue life in air to the formation of corrosion pits which presumably act as stress raisers, but this would not explain the conjoint action which is a feature of fretting fatigue.[17]

7.6 CONCLUSION

The effect of fretting on the fatigue strength of corrosion-resistant materials such as stainless steels and titanium alloys is very similar in an aqueous corrosive medium to that in air. This is ascribed to the chemical reactivity of the damaged surface, and is supported by the observation that fatigue life can be considerably prolonged by applying cathodic protection. Reducing the frequency of stressing reduces the fretting fatigue strength but this is thought to be the normal effect of frequency in corrosion fatigue.

REFERENCES

1. Heck, J. W. and Baker, E., *J. Soc. Naval Architects and Marine Engnrs.*, **71**, 327–46 (1963).
2. Battilana, R. E., *Chem. Engng.*, **78**, 130–2 (1971).
3. Smethurst, E. and Waterhouse, R. B., *J. Mat. Sci.*, **12**, 1781–92 (1977).
4. Bowden, F. P. and Tabor, D., *The friction and lubrication of solids*, Oxford University Press, pp. 155–6 (1950).
5. Ijzermans, A. B., *Wear*, **14**, 397–404 (1969).
6. Waterhouse, R. B., *Tribology*, **3**, 158–62 (1970).
7. Waterhouse, R. B., Dutta, M. K. and Swallow, P. J., *Mechanical behaviour of materials*, The Soc. of Mat. Sci., Japan, 292–8 (1972).
8. Taylor, D. E., and Waterhouse, R. B., *Corr. Sci.*, **14**, 111–22 (1974).
9. Waterhouse, R. B. and Dutta, M. K., *Wear*, **25**, 171–5 (1973).
10. Waterhouse, R. B., *Fretting corrosion*, Pergamon, Oxford, p. 188 (1972).
11. Higham, P. A., Stott, F. H. and Bethune, B., *Wear*, **47**, 71–80 (1978).
12. Higham, P. A., Stott, F. H. and Bethune, B., *Corr. Sci.*, **18**, 3–13 (1978).
13. Waterhouse, R. B. and Lamb, M., *Wear*, **60**, 357–68 (1980).

14. Waterhouse, R. B. and Taylor, D. E., Fundamental aspects of fretting corrosion. Final Report on SRC Contract B/SR/7946, University of Nottingham, 2 Feb. 1973.
15. Fenner, A. J. and Field, J. E., *Trans. N.E. Coast Instn. Engrs. Shipbldrs.*, **76**, 184–228 (1960).
16. Poon, C. and Hoeppner, D. W., *Wear*, **52**, 175–91 (1979).
17. Waterhouse, R. B., Wear, **58**, 391–2 (1980).

8

Fretting Fatigue in High Temperature Oxidising Gases

D. E. TAYLOR

Wolfson Institute of Interfacial Technology, University of Nottingham, UK†

8.1 INTRODUCTION

There are many practical applications in which moving components of machines are required to function at elevated temperatures. In certain cases, particularly in the gas turbine and power generation fields, it is necessary for components to operate in contact at high temperatures with the possibility of planned or fortuitous sliding occurring between the contacting surfaces. Under these circumstances the removal of material from the mating faces, i.e. wear, is likely to occur. Where the movement is oscillatory and of small amplitude, generally less than 75 μm,[1] the process is termed fretting. The movement may arise from unavoidable machine vibration, or from the cyclic stressing of one of the components, in which case the more serious consequence is the possible initiation and propagation of fatigue cracks at alternating stresses far below those at which such cracks would be initiated in the absence of fretting. Fretting fatigue can inflict damage consistent with strength reduction factors as high as ten.

Many design situations in turbine engines dictate that surfaces in loaded contact must move slightly relative to one another as a result of vibratory and other forces, and in operation various components such as the blades, the seat of the disc and the shaft are subject to cyclic stressing. There is in addition the hazard of elevated temperatures superposed on the fretting problem which may affect not only the mechanical properties of the materials but also enhance the effects of the environment, leading to

† Present address: Department of Materials and Production Engineering, National Institute of Higher Education, Limerick, Ireland.

oxidation and other forms of high temperature corrosion and the possibility of corrosion fatigue.

It has long been realised that the surface oxide films that form naturally on metals can be very effective in reducing wear between contacting metal surfaces. The success of new, high temperature oxidation resistant alloys depends on the spontaneous generation of a stable protective surface oxide film of limiting thickness. At elevated temperatures in oxidising environments, films on metal surfaces grow to greater thicknesses and generally retain good protective properties where they are not unduly stressed. However, any process which disrupts the integrity of the protective layer may have harmful consequences; the mechanical action of fretting is one such process.

A survey of published literature has indicated that little work has been carried out on the effects of fretting on high temperature fatigue although it has been studied as a wear problem. There is evidence to suggest that a number of metals and alloys exhibit a decrease in wear rate once the temperature is raised beyond a threshold value. Wright[2,3] proposed that above a transition temperature, specific for a given metal, rapid oxidation kinetics might promote enhanced fretting wear resistance. Bill[4] found that raising the temperature from 23 to 540 °C produced a marked drop in the fretting wear of nickel-based alloys with the wear rate remaining constant at a low value between 540 and 816 °C. Similar results were obtained by Hurricks,[5–7] who by increasing the temperature reduced the fretting wear of mild steel in an oxidising environment quite dramatically. In general terms, interpretations of these latter temperature effects invoked the concept of a critical thickness of oxide which needs to be maintained despite the disruptive nature of fretting. The rate of repair of the oxide is thus important and this depends on such conditions as temperature, frequency of the fretting motion and the nature of the environment.

Thus when fretting and oxidation occur simultaneously there is every likelihood that the process of oxide formation will be affected by the fretting action. Equally, the nature of the oxidation product can itself have a marked influence on the degree and type of damage produced during fretting. Several effects are possible including disruption of the oxide film producing abrasive debris which accelerates wear, or indeed complete removal of the oxide allowing metal-to-metal contact, local adhesion, possible seizure and surface gouging with subsequent effects on fatigue life. There is also the possibility at high temperatures of the so-called 'glaze' oxide formation as observed in the wear tests of Wood and co-workers.[8,9] They confirmed that in certain alloy systems the coefficient of friction and wear under conditions of reciprocating sliding fell as the temperature was raised, the 'glaze' oxide perhaps acting as a solid lubricant type layer and affording some protection against wear.

However, as previously outlined, one of the most damaging aspects of fretting is the large reductions which are experienced in fatigue strength. Room temperature studies of a wide variety of alloys are abundant but to date there has been little published information on the fatigue behaviour of materials at high temperatures in the presence of fretting. That which has been reported generally concerns practical attempts to reduce the effects of fretting fatigue at high temperatures and usually involves some form of surface treatment, including hardening of the surfaces or the application of metallic or cermet wear-resistant coatings. The relative efficiency of such methods will be reviewed in this chapter with reference to titanium, nickel, aluminium and iron-based alloy systems.

8.2 TITANIUM-BASED ALLOYS

Titanium and its alloys present one of the most critical challenges in materials engineering for modern gas turbine engines and aircraft structures. Whilst it offers excellent strength-to-weight advantages and other favourable properties, its fatigue life is extremely sensitive to surface conditions, particularly to damage from wear. Gas turbines represent an intensive commitment to titanium alloys, their effective performance depending on methods developed to protect the metal from wear damage. Several workers have chosen to investigate the effects on fatigue life of fretting titanium alloys at elevated temperatures.[10–17] Attempts have also been made to evaluate the effectiveness of a variety of surface treatments and coatings.

Betts'[10] work was concerned with fatigue in the presence of fretting around large bolt holes of aircraft compressor disc assemblies. One aim of his research was to further the study of the effects of elevated temperature on the fretting fatigue properties of uncoated, forged shot peened Ti–6Al–4V. Specimens were subjected to high cycle bending fatigue under a mean tensile stress with wear shoes clamped to the gauge length of the specimens. The fatigue bending stresses produced minute alternating strain motion of the specimens against the wear shoes causing *in situ* fretting damage. The severity of fretting as influenced by metal temperature was measured by change in fatigue life compared to non-fretted specimens. Three test temperatures, 20, 200 and 340 °C, were investigated at a mean stress of 140 MN m^{-2} and contact pressure of 175 MN m^{-2} (Fig. 8.1). Room temperature fretting was found to have little effect on the fatigue strength at 10^7 cycles. Fretting at 200 °C lowered the fatigue strength by approximately 15%; furthermore, the fretting fatigue life in the overstress region (70 MN m^{-2} above the non-fretted run out stress) was lowered by two orders of magnitude compared to results in the absence of fretting. At

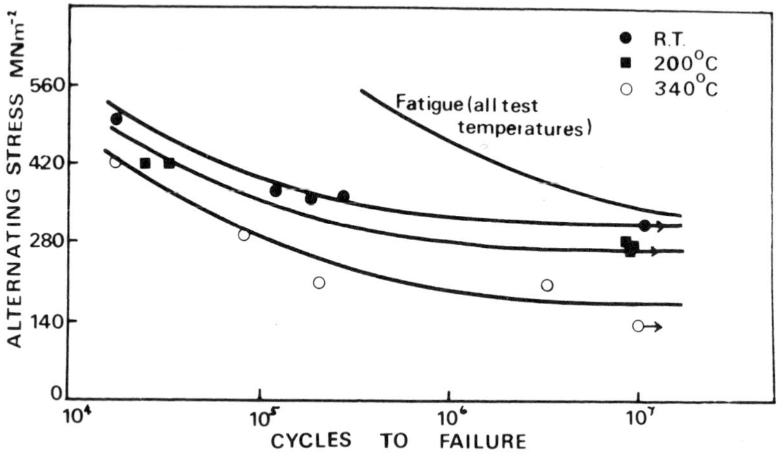

FIG. 8.1. Effect of simultaneous fretting and fatigue of shot-peened Ti–6Al–4V at room temperature, 200 and 340 °C at a mean tensile stress of 140 MN m^{-2}. (After reference 10.)

340 °C, fretting similarly reduced specimen life at overstress levels; however, more importantly, the fretting fatigue strength at 10^7 cycles was reduced to approximately 40% of that found under room temperature conditions.

The gross result of fretting normally is fatigue failure brought about by surface damage in conjunction with normal or transient high stresses in a component. It should be said at the outset that visual assessment of fretting mildness or severity is inconclusive by itself, in that the presence of more or less fretting debris on a microscopic examination is not necessarily relevant to the loss of surface integrity. It may in fact be misleading and should not be relied upon for assessing the severity of fatigue life degradation. It is the stress state acting in concert with stress raisers (e.g. pits, tears, cracks) which determine the actual fatigue propensity.

Returning to Betts'[10] work, he also examined the surfaces of his fretted specimens. Adhesive build-up of debris and gross pitting characterised the elevated temperature tests, the damage being more severe at 340 °C. Material adhering to the specimens corresponded to that removed from the wear shoe surfaces. In both room and elevated temperature tests the debris consisted largely of titanium oxide (rutile TiO_2) with fragments of unoxidised metal also being evident. Betts suggested that accumulation of the wear debris separates the metal surfaces causing the wear to intensify at the remaining areas of contact leading to the formation of the observed pits and contributing to the locally severe damage.

Betts[11] followed up the above research with attempts to identify, evaluate and characterise protective treatments for the alleviation of fretting wear on the fatigue life of titanium alloys to be employed at elevated temperatures. Several treatments including a baked-on aluminium pigmented film, a surface hardening cyaniding salt bath process, a shim insert of a low friction aluminium–bronze alloy and a shim of plain copper foil, were initially selected for evaluation at room temperature. Each treatment was applied to both fatigue specimen and wear shoes and protection assessed by the cyclic life achieved at an alternating bending stress of 410 $MN\,m^{-2}$ (overstress region). For various reasons, the first three processes referred to above proved totally ineffective under these room temperature conditions. The plain copper shims (0·25 mm thick) were by far the most protective, effectively inhibiting fretting fatigue of the Ti–6Al–4V alloy. The copper shims were observed to fret sacrificially forming a protective layer of copper debris. This layer accommodated the strain motion, was thought by Betts to redistribute the contact forces, did not itself abrade the alloy substrate, and was sufficiently cohesive to be retained within the interfaces. The copper shim technique was further tested in fretting fatigue at 340 °C where it was again found to inhibit fretting of the Ti–6Al–4V substrate, there being no evidence of the surface pitting such as occurred with the uncoated alloy. The copper shim insert, although found to be only partially effective in preventing fretting fatigue of another titanium alloy, Ti–5Al–4Mo–4Cr–2Sn–2Zr (Ti–17), provided at 340 °C considerably better behaviour than the non-shimmed system. While the surface damage of this alloy was negligible it appeared to be more sensitive than Ti–6Al–4V to contact stress raisers resulting from debris accumulating in areas beneath the wear shoes.

Syers and Harris[12–15] have carried out experimental studies to evaluate the resistance to fretting fatigue at room temperature, 350 and 500 °C, of a wide variety of surface treatments applied to the titanium alloys, Ti–6Al–4V and Ti–6Al–5Zr.

All specimens were tested in a repeated tensile mode. Fretting fatigue was accomplished by using a 'loading ring–fretting pad' assembly to apply a contact pressure of 93 MN m $^{-2}$ to the surface of the test pieces, a pressure which in most cases inhibited the relative slip amplitude over the majority of the pad area. Fatigue nucleation usually occurred at the slip-no-slip boundary in a fretting wear damaged area. The experimental method was regarded as severe, the fretting pad faces being untreated in all cases. The majority of room temperature tests were carried out in Amsler Vibrophores whilst a slipping-clutch resonant-spring machine was employed, at a load cycle frequency of 38 Hz, for the tests at elevated temperatures.

The process of shot peening was found by Syers and Harris to play an important role in protecting against fretting fatigue. A cold working technique such as shot peening is capable of inducing residual compressive stresses in the worked surface approaching yield strength values. The superposition of residual compression on a metallic surface subject to fretting may lead to a mitigation of the damaging tensile stress fields with a significant improvement in fatigue strength. The beneficial effect of shot peening on the alleviation of fretting fatigue strength is demonstrated for Ti–6Al–4V and Ti–6Al–5Zr by the results presented in Table 8.1.

Fretting of plain polished specimens at both test temperatures caused a lowering of the room temperature fatigue strength by almost 80%. Shot peening of Ti–6Al–4V led to a small reduction (7%) in the room temperature fatigue properties and a similar reduction in the 350 °C fatigue strength. However, referring to Table 8.1 it will be seen that at both room temperature and 350 °C, fretting fatigue strengths of the shot-peened alloy were more than twice those found for unpeened specimens. The results of shot peening Ti–6Al–5Zr to Almen A16 were also encouraging. The treatment actually improved the room temperature fatigue properties by 14% and had no deleterious effect on the fatigue strength at 500 °C. Resistance to fretting fatigue at both room temperature and 500 °C was again markedly better than that found for the untreated alloy.

Syers also investigated the relative fretting resistance of various coatings applied by detonation and diffusion bonding techniques. The results obtained from three particular coatings are presented in Table 8.2 together with those for the untreated Ti–6Al–4V alloy.

The application by a detonation process of tungsten carbide +15% cobalt produced a coating exhibiting excellent adhesion and yielded promising results providing the alloy was shot peened prior to coating. The reductions of fatigue resistance due to fretting were less than those experienced by the untreated material at both room temperature and 350 °C. Although coating the alloy lowered the 350 °C fatigue strength by

TABLE 8.1

EFFECT OF SHOT PEENING ON FATIGUE AND FRETTING FATIGUE OF Ti–6Al–4V AND Ti–6Al–5Zr

Material	*Specimen treatment*	*Temperature*	*Fatigue strength ($MN\,m^{-2}$) 10^7 cycles*		*Percentage reduction of basic R.T. properties*	
			Unfretted	*Fretted*	*Unfretted*	*Fretted*
Ti–6Al–4V	Plain polished	R.T.	650	140	0	78
	Plain polished	350 °C	580	140	11	78
	Shot-peened Almen A7	R.T.	600	340	7	48
	Shot-peened Almen A7	350 °C	540	310	17	52
Ti–6Al–5Zr	None (as received)	R.T.	490	125	0	74
	None (as received)	500 °C	430	200	12	59
	Shot-peened Almen A16	R.T.	560	250	(−14)	49
	Shot-peened Almen A16	500 °C	430	260	12	47

TABLE 8.2

EFFECT OF DETONATED AND DIFFUSION BONDED COATINGS ON FATIGUE AND FRETTING FATIGUE OF Ti–6Al–4V

Specimen treatment	*Temperature*	*Fatigue strength ($MN\,m^{-2}$) 10^7 cycles*		*Percentage reduction of untreated R.T. fatigue properties*		*Percentage reduction of untreated 350 °C fatigue properties*	
		Unfretted	*Fretted*	*Unfretted*	*Fretted*	*Unfretted*	*Fretted*
None (as received)	R.T.	700	230	0	67	—	—
	350 °C	540	190	23	73	0	65
Detonation process: (WC + 15 % Co) after shot peening to A16	R.T.	550	380	21	46	—	—
	350 °C	370	280	47	60	31	48
Detonation process: (Cu–Ni–In) after shot peening to A8–12	R.T.	520	340	26	51	—	—
	350 °C	500	225	29	68	7	58
Diffusion bonding: Cr plate + D.B.	R.T.	230	140	67	80	—	—
	350 °C	150	not tested	79	—	72	—

approximately 30 %, fretting fatigue resistance at this temperature showed a significant improvement (47 %) over that exhibited by the untreated material. The diffusion of chromium into the surface of the titanium alloy was not so successful; the application (to a coating thickness of 0·038 mm) caused a catastrophic lowering of fatigue strength by 67 %; a further 13 % reduction resulted from fretting at room temperature. However, as can be seen from the histogram, Fig. 8.2, this was remedied by diffusion bonding a thin layer of chromium, shot peening, and then finally chromium plating. Following this treatment, the fatigue and fretting fatigue strengths were 93 % and 51 % respectively of the strength of the untreated alloy fatigued at 350 °C.

Coating the specimens after shot peening with a layer of copper–nickel–indium led to a reduction of the room temperature fatigue resistance by 26 %. This treatment had little adverse effect (7 % reduction) on the 350 °C fatigue strength. However, under fretting fatigue conditions, the coating tended to extrude from areas of major asperity contact beneath the fretting pads, leaving islands of parent material which became sites for crack nucleation. Although the fretting fatigue strength at 350 °C was an 18 % improvement on that for the untreated alloy this amounted to only 32 % of the basic room temperature fatigue strength.

The effectiveness of several other anti-fret treatments evaluated by Syers at 350 °C have been compiled into a histogram (Fig. 8.2). Of these the best resistance to fretting fatigue was provided by shot peening to an intensity of Almen A16 followed by the application of a high velocity plasma flame-sprayed coating of tungsten carbide + 12 % cobalt. This procedure produced little reduction in the fatigue properties and 56 % of basic properties in fretting fatigue. Some problems were encountered in attaining good adhesion of this coating, but even when the coating was only held in place by the fretting pads, good protection was generally afforded. A diffusion-bonded electrodeposit of nickel subsequently shot peened to Almen A7 again had little adverse effect on fatigue and gave reasonable protection against fretting fatigue as did the aforementioned diffusion-bonded electroplated chromium which was shot peened and then replated.

Surface anodising techniques were also evaluated at 350 °C. A treatment based on sulphuric acid anodising produced hard (but thin) films on the surface of the alloy. Such anodic films, *per se*, or when used as a basis for dry film lubricants bonded with epoxy or polyimide resins provided little, if any, protection against fretting fatigue. It appeared that the acid anodic films 'craze-cracked' and 'spalled' under the fretting pads. Softer (thicker) coatings as provided by the 'Watervliet' process (essentially high current

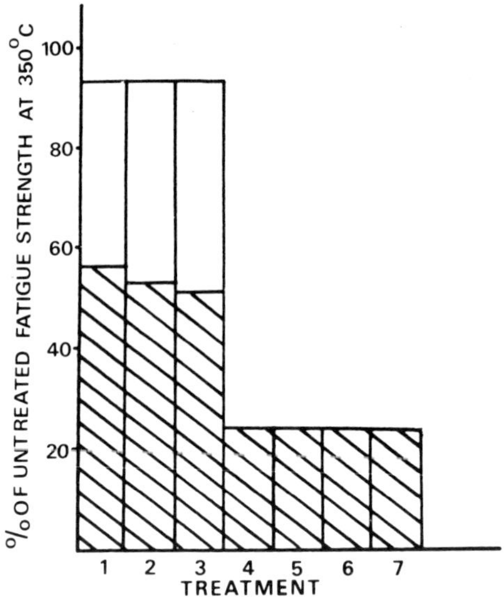

FIG. 8.2. A summary of effects of some surface treatments on the fatigue, □, and fretting fatigue, ▧, properties of Ti–6Al–4V at 350 °C.

1. Shot-peened Almen A16. High velocity plasma flame-sprayed WC + 12% Co.
2. Electroplated Ni. Diffusion bonded + shot-peened Al A7.
3. Electroplated Cr. Diffusion bonded + shot-peened Al A8-12 + electroplated Cr.
4. Hard anodised.
5. Hard anodised + (epoxy resin + MoS_2).
6. Watervliet anodised + (polyimide resin + graphite).
7. Watervliet anodised + (epoxy resin + MoS_2).

(After reference 15.)

density anodising in sodium silicate–sodium hydroxide) also proved of little use as anti-fret protection, the poor adherence of the loose powdery films leading to early detachment from the titanium surfaces.

The extensive work of Syers and Harris on titanium alloys has shown that, in general, of the coatings examined, the best results are given by the hard metal deposits applied by detonation gun, plasma flame-spray or electrodeposition techniques. It is also clear that the shot peening process offers good protection against fretting fatigue, and it appears to be a necessary pre- or post-treatment for the more successful anti-fret coatings, mainly to offset the adverse effects that the coatings may have on the basic fatigue properties.

Pavliscak[16] is another researcher who has studied the fretting fatigue behaviour of annealed Ti–6Al–4V at elevated temperatures and as with the previous workers cited has investigated the palliative effects of shot peening and surface coating. Fretting and fatigue were combined by use of a rotating cantilever beam machine modified to incorporate the fretting device. The fatigue specimen consisted of a straight rod tapered at one end to mate with a matching tapered hole in a cylindrical fretting ring, the distance the ring was pressed on to the specimen controlling the initial interface pressure. With this arrangement a bending load applied to the specimen caused slight axial movement of the specimen relative to the fretting ring, the load becoming cyclic as the entire assembly rotated. Both specimen and fretting device were enclosed in a furnace.

As shown in Table 8.3, and as in line with the results of Syers[15] presented in Table 8.1, Pavliscak found that raising the temperature from 20 to 400 °C had no adverse effect on the fretting fatigue strength of Ti–6Al–4V fretted against like material at a normal pressure of 35 MN m^{-2}. On increasing this pressure to 140 MN m^{-2} the fretting fatigue strength at 400 °C was reduced by 12% compared to that found at room temperature. Shot peening was seen to increase the fretting fatigue strength, the improvement being greater at room temperature than at 400 °C, again a trend shown in Syers' results.

The application of a plasma spray coating of copper–nickel–indium (150 μm thick) to the shot-peened specimen had little further beneficial effect on the fretting fatigue strength at 400 °C but at room temperature led to a significant increase in the shot-peened value, raising it from 270 to 320 MN m^{-2} which represented a 45% improvement on the untreated value. Pavliscak also looked into the influence of material combination on the fatigue life of the titanium alloy by fretting it against a 4340 type medium carbon alloy steel. Table 8·3 indicates that at both 20 and 400 °C this resulted in a small improvement in fatigue resistance relative to that for titanium against titanium, with shot peening again leading to an increased fretting fatigue strength at both test temperatures. As with the titanium–titanium combination, coating of the shot-peened specimens with copper–nickel–indium greatly improved the fretting fatigue strength at 20 °C, an improvement attributed to sacrificial wearing of the coating (as with the copper shims employed by Betts[11]). At 400 °C the effect of the coating was lost, the strength falling to a value approaching that for the untreated alloy.

Thus, except for the untreated alloy, testing at 400 °C led to a lowering of fretting fatigue strengths compared to those found at 20 °C, the greatest reductions in fact being associated with the copper–nickel–indium treatment which provided the best fretting fatigue resistance at room

TABLE 8.3

EFFECT OF SHOT PEENING, COATING AND MATERIAL COMBINATION ON THE FRETTING FATIGUE PROPERTIES OF Ti–6Al–4V AT 20 AND 400 °C

Fretting ring material	*Ti–6Al–4V fatigue specimen surface treatment*	*Normal pressure* ($MN\,m^{-2}$)	*Fretting fatigue strength* ($MN\,m^{-2}$) *3 × 10⁶ cycles*		*Percentage reduction at 400 °C*
			20 °C	*400 °C*	
Ti–6Al–4V	None (as received)	35	220	220	0
	Shot peened	35	270	240	11
	Shot peened + Cu–Ni–In	35	320	245	23
	None	140	215	190	12
4340 steel	None (as received)	35	230	250	(−9)
	Shot peened	35	290	290	0
	Shot peened + Cu–Ni–In	35	325	255	22

temperature. Note, however, that at 400 °C both treatments employed by Pavliscak proved more resistant to fretting than did the untreated alloy surface.

In a more recent study, Hamdy and Waterhouse[17] have investigated the fretting fatigue behaviour of mill-annealed Ti–6Al–4V at temperatures up to 600 °C. Fretting was produced by clamping bridges on to the gauge length of a fatigue specimen with a loading ring, the fatigue being provided by a small push–pull machine operated in fluctuating tension at a frequency of 50 Hz. Fatigue curves were determined in the presence of fretting at 20, 200, 400 and 600 °C with the specimen and bridges surrounded by an independently supported split furnace as shown in Fig. 8.3. Testing was carried out at a mean stress of 250 MN m^{-2} and a clamping pressure of 32 MN m^{-2}, both of which were adjusted after the specimen had reached the required temperature. The fretting fatigue results are shown in Fig. 8.4 together with a fatigue curve for the alloy at 600 °C.

FIG. 8.3. The experimental arrangement for determining fatigue curves in the presence of fretting at elevated temperatures. The top half of the furnace has been removed showing the specimen and fretting bridges held by a proving ring. (After reference 17.)

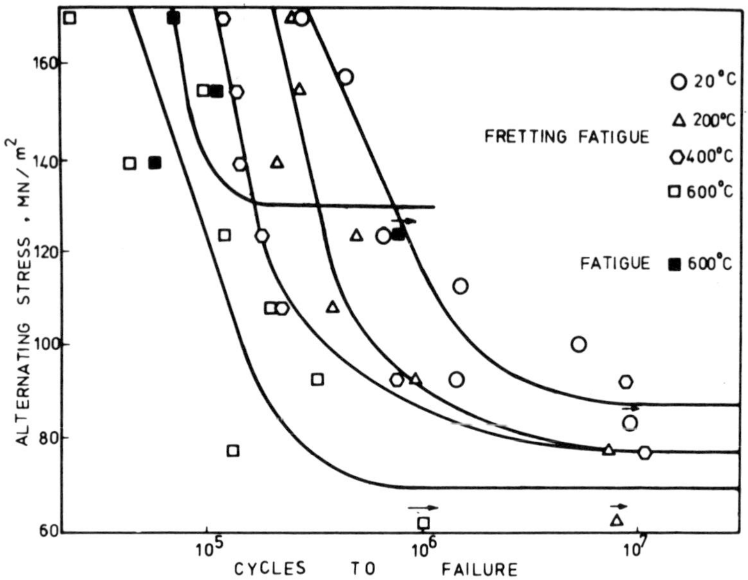

FIG. 8.4. Fatigue curves for Ti–6Al–4V in the presence of fretting at a mean stress of 250 MN m^{-2} at 20, 200, 400 and 600 °C together with the fatigue curve at 600 °C. (After reference 17.)

The already low fretting fatigue strength at room temperature decreased still further, but not dramatically as the temperature was successively raised to 200, 400 and 600 °C. However, as Fig. 8.4 indicates, there was a significant decrease in fatigue life at stresses above the fatigue limit with increase in temperature. This is attributed to the deterioration in fatigue properties at high temperatures, as illustrated by the short lives experienced by the specimens tested in plain fatigue at 600 °C, rather than any specific aggravation of the fretting damage. The general impression obtained from examination of the fretted surfaces was of transfer of material from one surface to the other which became much greater in volume as the temperature was raised. In several cases as noted in Betts' work[10] previously discussed, material adhering to the fatigue specimen corresponded very closely to that removed from the bridge foot.

The authors suggest that failure of the specimens was associated with the formation of ridges on the surface which developed into piles of platelets as the fretting action continued (Fig. 8.5). In some instances, particularly at 600 °C, these accretions appear to be firmly stuck together growing to large proportions, individual platelets having a reproducible shape as seen in Fig. 8.6.

FIG. 8.5. Piled up platelets on a fretted surface in the vicinity of the site of the fatigue fracture of a Ti–6Al–4V specimen fretted at 200 °C. (After reference 17.)

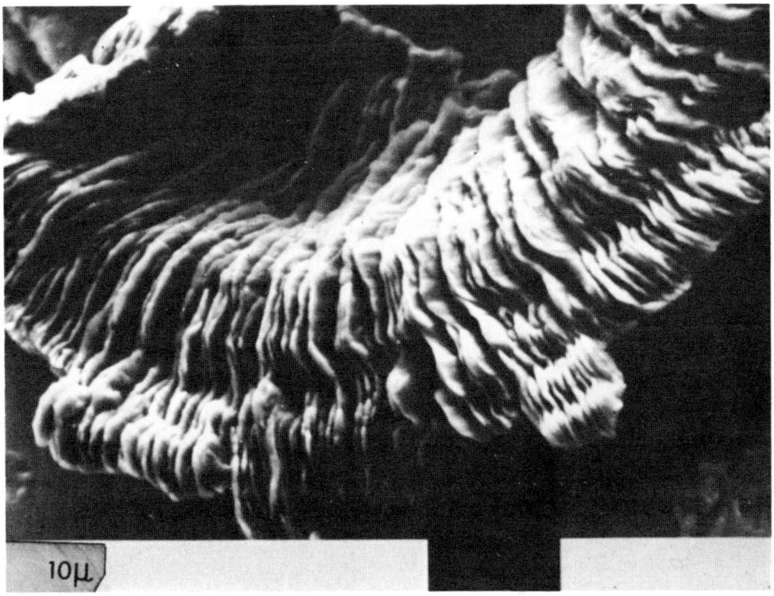

FIG. 8.6. Ti–6Al–4V fretted at 600 °C. Individual platelets of apparently reproducible shape piled up into accretions. (After reference 17.)

8.3 ALUMINIUM-BASED ALLOYS

Harris[13,14,18–20] has reported research, the object of which was to find means of protecting aluminium alloys against fretting fatigue at temperatures up to 150 °C and so alleviate problems experienced in Mach 2 supersonic transport aircraft and engine compressor components. Prerequisites for an anti-fret medium for use at such temperatures involves consideration of several factors, which include chemical compatibility with the parent alloy, permanency *in situ* and flow resistance at elevated temperature. Any treatment must also have no deleterious effect upon the fatigue life of the alloy to which it is being applied. Harris explored several promising avenues based mainly on the application of solid lubricants, such as molybdenum disulphide (MoS_2), incorporated in a vehicle with the requisite rheological properties.

The fretting fatigue tests were carried out on sheet specimens against which two pads of steel were held at a pressure of 46 MN m^{-2} by a loading ring. The assembly was placed in a split furnace and fatigue tested in a slipping clutch machine with fatigue loading by repeated tension at a frequency of 38 Hz. Harris tested the behaviour of two aluminium alloys. The first, containing 2·5 % copper and clad with a 1 % zinc–aluminium alloy (DTD 5070A), and the second an unclad alloy with 2 % copper (RR58). The clad alloy was tested in the 'as rolled' condition and the unclad polished to a 10μm finish. Anti-fretting coatings applied to both sides of the alloy specimens were fully cured, or allowed to cure at room temperature as appropriate.

In all, fifteen coatings were tested by Harris. The results from the three best treatments have been selected for presentation in Table 8.4. Only two coatings completely eliminated the effects of fretting; the first was basically an epoxy resin containing MoS_2 which required curing at 150 °C; the second, the only one of many air-drying compounds to be successful, was an isocyanate cured epoxy resin loaded with MoS_2 which was equally successful in negating the effects of fretting on both the clad and unclad alloys. A carbon-fibre resin film (Hyfil) was also found to eliminate fretting completely on the unclad alloy. Chromic acid anodising and a flame-sprayed stainless steel matrix loaded with sintered PTFE were among the treatments which proved largely ineffective.

Syers[21] furthered the work of Harris by evaluating the anti-fret properties of MoS_2 in the isocyanate cured epoxy resin at temperatures up to 250 °C. In tests on the clad alloy in an Amsler Vibrophore at a frequency of 47 Hz, Syers confirmed that the treatment was completely effective at

TABLE 8.4

THE EFFECT OF ANTI-FRET TREATMENTS ON THE FRETTING FATIGUE STRENGTH OF ALUMINIUM ALLOYS AT 150 °C

Specimen material	*Specimen treatment*	*Fatigue strength ($MN m^{-2}$) 10^7 cycles*		*Percentage reduction on fretting*
		Unfretted	*Fretted*	
DTD 5070A (clad)	None (as received)	90	45	50
	Epoxy resin $+MoS_2$	—	90	0
	Isocyanate cured epoxy resin $+MoS_2$	—	90	0
	Thickened petroleum oil $+MoS_2$	—	75	17
RR 58 (unclad)	None (as received)	170	56	67
	Isocyanate cured epoxy resin $+MoS_2$	—	170	0
	Carbon-fibre resin film	—	170	0

both room temperature and 150 °C. At 200 °C the coating afforded a degree of protection against fretting, the resistance being some 22 % better than that found for uncoated specimens. However, at 250 °C the coating proved totally ineffective, a value of fretting fatigue strength identical to that for the untreated alloy being recorded.

8.4 NICKEL-BASED ALLOYS

In gas turbine engines the blades are generally attached to the rotor discs by dovetail or fir-tree roots. Vibration in a rotating disc is a common experience and fretting at the fixing positions of the turbine blades is not unknown. In certain engine designs, the disc is bolted to the main shaft via a flange and fretting is possible under the bolt heads. Initiation of fatigue cracks in the disc at these areas can lead to the disc's catastrophic failure. The use of mechanically fastened turbine wheel assemblies also creates the possibility of motion between component parts and hence a reduction of fatigue life through fretting. Room temperature studies of various nickel-based alloys are abundant, but there is a general absence of fretting fatigue data at the elevated temperatures pertinent to the operation of gas turbines.

With these considerations in mind, Hamdy and Waterhouse[22] have investigated the fretting fatigue behaviour of a nickel-based alloy containing 19% chromium and 18% iron (Inconel 718) which has good corrosion resistance, strength and creep resistance at high temperatures, and is representative of the type of material used in gas turbines. Employing the experimental arrangement previously described[17] they determined fatigue curves with and without fretting (Fig. 8.7) at temperatures of 20, 280 and 540 °C at a frequency of 50 Hz, a mean stress of 550 MN m^{-2} and bridge clamping pressure of 32 MN m^{-2}. Tests were also attempted at 700 °C but problems were experienced due to creep at the higher alternating stresses. Although fatigue results at elevated temperatures were somewhat scattered they appeared generally to lie on one curve which exhibited longer fatigue lives at lower stress levels than did the room temperature curve.

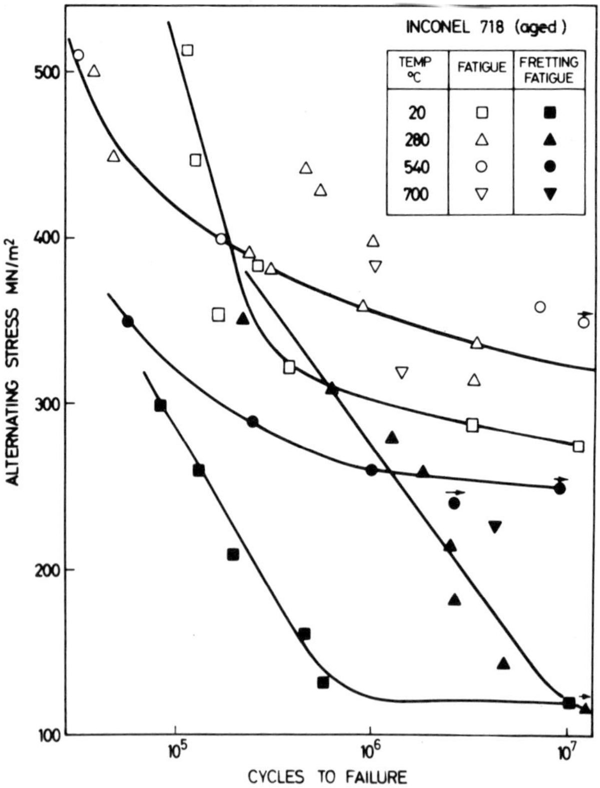

FIG. 8.7. Fatigue and fretting fatigue curves for Inconel 718 at an applied mean stress of 550 MN m^{-2}. (After reference 22.)

Fretting at room temperature and 280 °C produced identical fatigue strengths at 10^7 cycles but at higher stress levels the lives at 280 °C were considerably longer. At 540 °C the fretting fatigue strength showed a dramatic rise approaching the fatigue strength at room temperature in the absence of fretting. This rise the authors attributed to the formation of a protective oxide 'glaze'. As referrred to in the introduction to this chapter such glazes have been observed in the sliding wear of Nimonic and Incoloy type alloys in the temperature region 150–800 °C.[9] The formation of the glaze is characterised by a reduction in the coefficient of friction and a corresponding lowering of the wear rate, its appearance being described as showing 'many distinct, well-formed, small abrasion grooves, parallel to the direction of sliding'. The formation of the glaze is thought[8] to be due to the contact pressure and high localised temperatures in the load bearing regions causing thermal softening at points of contact between oxide particles resulting in a sintered compact mass. Hamdy and Waterhouse suggest that at 280 °C the glaze layer was formed but showed signs of breakdown by delamination (Fig. 8.8) and was unable to reform rapidly enough to keep pace with the fretting action. The full protective effect of the

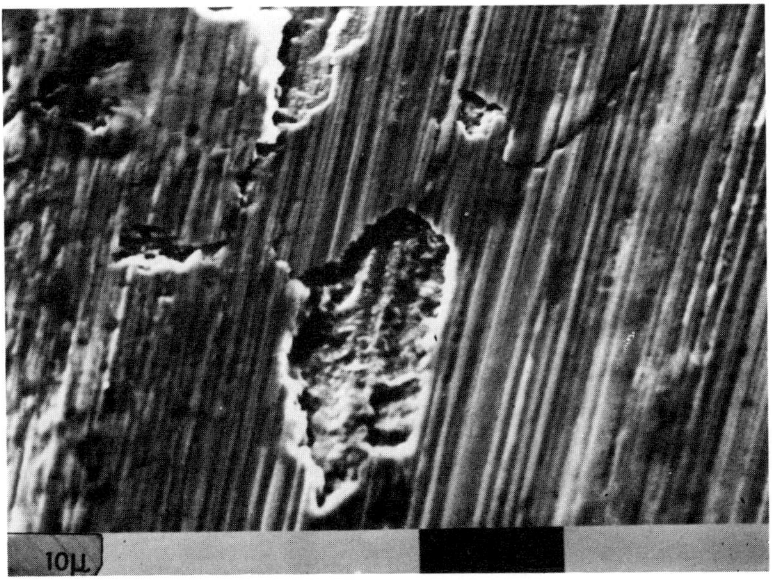

FIG. 8.8. 'Glaze' oxide disrupted by delamination during fretting of Inconel 718 at 280 °C. (After reference 22.)

glaze was seen only at 540 °C where extensive flawless areas were produced on both contacting surfaces.

Fielder[23] also concerned himself with the influence of high temperatures on the fretting fatigue behaviour of gas turbine superalloys, in particular an alloy containing 46 % nickel, 27 % iron and 15 % chromium (D979). The use of pre-calibrated bolts to apply a contact pressure of 280 MN m^{-2} via bridge-type pads to a fatigue specimen reduced the size of the assembly sufficiently for it to be tested inside a standard resistance wound furnace. Fatigue loading at a frequency of 60 Hz and mean stress of 550 MN m^{-2} was accomplished by an hydraulic activator in a closed loop servo-controlled axial fatigue machine. The majority of Fielder's tests were carried out at 480 °C, a temperature representative of that of the hub of an operating turbine disc with the chosen alloy being fretted against pads of a medium carbon alloy steel, both surfaces being in a ground condition.

Fretting reduced the fatigue strength after 30×10^6 cycles at 480 °C from 260 to 100 MN m^{-2}. Following shot peening of the D979 to Almen A6-8 the fretting fatigue strength increased to 170 MN m^{-2}, an improvement of 40 % on the degradation due to fretting in the non-peened alloy. Fielder noted that fretting damage on the specimens consisted of depressions approximately 50 μm deep containing highly compacted oxide debris which had a polished appearance, a description which may intimate the early stages of formation of a glaze such as those reported in the work of Hamdy and Waterhouse[22] discussed above.

8.5 IRON-BASED ALLOYS

The wear behaviour of metallic components, particularly those of steel, can frequently be improved by the application of certain types of flame-sprayed surface coatings. Initially used mainly for the application of zinc and aluminium for corrosion protection, the flame-spraying process has been developed and is now capable of depositing a range of metal, ceramic and cermet coatings for a wide variety of applications. Of the available metal coatings it is well known that molybdenum has the advantage of good adhesion to steel surfaces as well as being thought to confer good wear resistance. It can also be sprayed to give a range of coating hardness which can be controlled by varying the oxygen content of the inflammable gas mixture. These properties and the fact that in many industrial machines fretting occurs at elevated temperatures led Overs, Harris and Waterhouse[24,25] to investigate coating with sprayed molybdenum as a possible palliative against fretting fatigue of steel at 300 °C.

Following grinding and a grit blasting process the gauge lengths of cold drawn 0·4% carbon steel (En 8) fatigue specimens were sprayed with molybdenum, final grinding reducing the effective coating thickness to 0·15 mm. Both 'hard' and 'soft' coatings were produced on different specimen surfaces, the hardness being regulated by the oxygen partial pressure in the combustible gas mixture. A 'soft' bonding undercoat was provided for the 'hard' coated molybdenum. Uncoated specimens were electropolished prior to test. Push–pull fatigue loading was carried out in fluctuating tension at a mean stress of 380 MN m^{-2}, fretting being provided by a proving ring–bridge assembly, at a clamping pressure of 32 MN m^{-2}. Table 8·5 shows the fatigue strengths (10^7 cycles) determined with and without fretting at 20 and 300 °C. Both types of coating had a deleterious effect on the room temperature fatigue behaviour of the steel, the 'soft' and 'hard' molybdenum lowering the fatigue strength by 24 and 18% respectively. At 300 °C the coatings reduced the fatigue strength by a further 7–9%. Uncoated specimens exhibited fatigue strength of 310 MN m^{-2} at both room temperature and 300 °C; however, under fretting conditions this value was considerably reduced. At 20 °C a fatigue limit was still observed but at 300 °C the authors reported a large reduction in fatigue strength (60%) over the whole stress range, together with the removal of the fatigue limit. Molybdenum in the 'hard' condition appeared almost insensitive to fretting at room temperature reducing the fatigue strength by only 10 MN m^{-2} whilst fretting the 'soft' coating produced a value below that for the uncoated steel. At 300 °C the fretting fatigue performances of both coatings were similar and superior to that of the uncoated steel improving its strength at 10^7 cycles by approximately 50%.

Examination of the room temperature-tested uncoated steel specimens showed evidence of loose wear particles and the breaking away of surface layers. No such large scale surface disruption was observed on the coated specimens and at 300 °C, the load bearing surfaces of both 'hard' and 'soft' coated samples were covered by a grey, compact, adherent film likened to a glaze. It is suggested that the resistance of a molybdenum oxide (MoO_2) dispersion in the molybdenum coatings would provide the initial transient oxide particles which grow and sinter together to form a glaze and in such circumstances the initial surface damaging processes of fretting may be partially or totally eliminated. Results obtained under fretting fatigue conditions with both 'hard' and 'soft' coatings suggest that the glaze reduces friction at the sliding interface and inhibits the type of surface damage which leads to the initiation of fatigue cracks. The better results obtained at 20 °C by the 'hard' compared to the 'soft' coatings are related by the authors

TABLE 8.5

THE EFFECT OF 'HARD' AND 'SOFT' MOLYBDENUM COATINGS ON FATIGUE AND FRETTING FATIGUE STRENGTH OF 0·4 % CARBON STEEL AT 20 AND 300 °C

Material	*Fatigue strength ($MN\,m^{-2}$) 10^7 cycles*				*Percentage reduction of electro-polished fatigue strength*			
	20 °C		*300 °C*		*20 °C*		*300 °C*	
	Unfretted	*Fretted*	*Unfretted*	*Fretted*	*Unfretted*	*Fretted*	*Unfretted*	*Fretted*
Electropolished En 8 steel	310	160	310	120	—	48	—	61
En 8 + 'soft' molybdenum coat	235	145	215	180	24	53	31	42
En 8 + 'hard' molybdenum coat	255	245	225	185	18	21	27	40

to the extent to which the glaze oxide forms, the glaze forming less readily on the 'soft' coating due to its lower volume fraction of dispersed MoO_2. At 300 °C the glaze oxide forms more quickly allowing both 'soft' and 'hard' coatings to exhibit similar fatigue strengths.

Overs and Harris are currently working on electric arc-sprayed composite coatings. These are laid down by spraying two types of alloy wire, one from the anode and the other from the cathode of the arc pistol, and with this arrangement a wide range of coating mixtures is available. Fretting fatigue tests have been carried out on a low alloy creep resistant steel sprayed using alloy wires to provide a four-component coating of nickel, aluminium, iron and chromium. At 20 °C the coating reduces the fatigue limit of the steel by at least 30 % but improves the fretting fatigue strength by 35 %. However, at 475 °C the coating was found to have no deleterious effect upon the fatigue strength of the steel and improves the fretting fatigue strength by 11 % at this elevated temperature. In work that is continuing, the authors have also found that the oxidation resistance is excellent compared to that of the untreated steel and the fretting surface damage is considerably reduced.

8.6 SUMMARY

The need for machine components to operate at elevated temperatures under conditions which could lead to their fretting one against another has prompted several researchers to investigate practical means of protecting such components against fretting and the associated degradation in fatigue life.

The temperature regime up to 150 °C is typical of supersonic air transport where the mechanical behaviour of the aluminium alloy skins for the wings and fuselage is crucial for the performance and, not least, the structural integrity of the aircraft. The inevitable intrusion of fretting increases the importance of processes designed to protect the structure. On simple rheological grounds conventional lubricants and greases offer very little as 'anti-fret' agents in such situations since adequate flow resistance and permanency *in situ* require at least some form of solid film lubricant stable throughout the temperature range. For aluminium alloys, bare and clad, fretting fatigue investigations[13,14,18,19] have demonstrated that up to 150 °C, the above problems have substantially been countered, an air curing isocyanate epoxy resin matrix loaded with MoS_2 providing full protection against fretting. This scheme continued to afford some protection at 200 °C but became totally ineffective at higher temperatures.

For applications up to 500 °C interest has centred on titanium and nickel-based alloys with their obvious attraction of high specific strength. Titanium alloys are particularly sensitive to fretting damage and require 'anti-fret' media with a superior temperature resistance to the loaded polymer films so successful on the aluminium alloys. It is evident that shot peening, capable of inducing residual compressive stresses into the surface material, substantially improves fatigue strength. Research has also shown that hard metal deposits including tungsten carbide–cobalt coatings, sprayed by detonation and high velocity plasma techniques, improve quite significantly the fretting fatigue strength of titanium alloys at elevated temperatures. Promise has also been shown by diffusion bonding of chromium and nickel,[12–15] although it has been clearly demonstrated that shot peening plays an important role in all the above mentioned processes essentially by offsetting the adverse effects the coatings may have on the basic fatigue properties.

Shot peening also proved effective in inhibiting fretting fatigue of nickel-based alloys at elevated temperatures. Fretting of a nickel-based alloy at 480 °C[23] reduced its fatigue strength by approximately 60 %; however, shot peening enabled more than 40 % of the degradation in fatigue life to be recovered.

The generation at elevated temperatures of a so-called 'glaze' oxide plays a significant role in the fretting fatigue behaviour of the aged nickel-based alloy Inconel 718.[21] As the temperature was raised from 280 to 540 °C the fretting fatigue strength increased by 130 % and was almost restored to the room temperature fatigue value in the absence of fretting. This marked improvement has been attributed to the development of a glaze oxide on the surface of the alloy. The surface itself was smooth and on examination showed many characteristic well-formed parallel markings produced by the sliding.

It is suggested that the formation of the glaze under the conjoint action of oxidation and sliding is due to the interfacial pressure and the high localised temperatures developed in the load-bearing regions. This causes thermal softening at points of contact between oxide particles resulting in a sintered compact mass. For the glaze to confer effective protection it must form rapidly with the onset of fretting, since it is known that fatigue cracks are initiated in the early stages of the process, it must not break down easily, and finally, it must be capable of rapidly repairing itself.

The efficiency of a glaze oxide as a fretting fatigue palliative has also been effectively demonstrated in tests carried out at temperatures up to 300 °C on a steel substrate coated with molybdenum.[24,25] If steel engineering components are to be operated under fretting fatigue conditions at elevated temperatures there are clear advantages to be gained by their coating with flame-sprayed molybdenum. Such coatings contain a fine dispersion of oxides which are thought to promote the formation of glazes which in turn reduce adhesion and friction at the sliding interfaces and lead to improved fretting fatigue resistance.

It has been demonstrated that the fretting fatigue behaviour at elevated temperatures of the various alloy systems discussed may be improved by selected methods of surface treatment. Shot peening, various types and techniques of coating, and the generation of surface glaze oxides have been shown under specified conditions to be particularly effective fretting mitigators.

REFERENCES

1. Waterhouse, R. B., *Fretting corrosion*, Pergamon Press, Oxford (1972).
2. Wright, K. H. R., National Engineering Laboratory Leaflet No. 394 (1962).
3. Wright, K. H. R., National Engineering Laboratory Report No. 402 (1969).
4. Bill, R. C., NASA Tech. Note TN D-7570 (1974).
5. Hurricks, P. L., *Wear*, **19**, 207–29 (1972).
6. Hurricks, P. L., *Wear*, **30**, 189–212 (1974).
7. Hurricks, P. L., *Industrial Lubrication and Tribology*, **26**, 209–14 (1975); **27**, 9–16 (1976).
8. Stott, F. H., Lin, D. S. and Wood, G. C., *Corros. Sci.*, **13**, 449–69 (1973).
9. Lin, D. S., Stott, F. H. and Wood, G. C., *ASLE Trans.*, **17**, 251–62 (1973).
10. Betts, R. K., Air Force Materials Laboratory Technical Report, AFML-TR-71-212, Wright-Patterson Air Force Base, Ohio (1971).
11. Betts, R. K., Air Force Materials Laboratory Technical Report AFML-TR-74-18, Wright-Patterson Air Force Base, Ohio (1974).
12. Harris, W. J., Rolls-Royce Research Report R245/52/69 (1969).
13. Harris, W. J., AGARD Advisory Report 45 (1972).
14. Harris, W. J. *Fretting in aircraft systems*, AGARD Conference Proceedings No. 161, 1974, AGARD (1975).
15. Syers, G., Rolls-Royce Research Report LLR 10245/FS/78 (1978).
16. Pavliscak, T. J., Ph.D. Thesis, The Ohio State University (1968).
17. Hamdy, M. M. and Waterhouse, R. B., *Wear*, **56**, 1–8 (1979).
18. Harris, W. J., AGARD Advisory Report 8 (1967).
19. Harris, W. J., AGARD Advisory Report 21 (1970).
20. Harris, W. J., Ministry of Defence D. Mat. Report No. 174 (1971).
21. Syers, G., Rolls-Royce Research Report LLR 10251 (1976).
22. Hamdy, M. M. and Waterhouse, R. B., *Wear of materials*, ASME, New York, 351–5 (1979).

23. Fielder, L. J., Materials and Process Technology Laboratories, Report MR 0326, Avco Lycoming Division, Stratford, Connecticut (1976).
24. Overs, M. P., Harris, S. J. and Waterhouse, R. B., *Wear of materials*, New York, ASME, 379–87 (1979).
25. Overs, M. P., Harris, S. J. and Waterhouse, R. B., *Trans. Inst. Metal Finishing*, **57** 81–8 (1979).

9

Theories of Fretting Processes

R. B. WATERHOUSE

Department of Metallurgy and Materials Science, University of Nottingham, UK

9.1 INTRODUCTION

Fretting involves a number of factors such as contact between surfaces, frictional effects, adhesion, wear, surface fatigue, stress concentrations, corrosion and environmental effects, all of which are major subjects in their own right. This complexity is a possible reason for the comparatively few investigations into fretting and fretting fatigue.

It is generally accepted that one of the major factors in the initiation of fatigue cracks by fretting is the alternating shear stress in the surface arising from the friction force occasioned by the oscillatory movement. Nishioka and Hirakawa derived an expression for the alternating stress σ_{fw1} to initiate a fatigue crack under fretting conditions for the case of a cylindrical pad in contact with the plane surface of a fatigue specimen:[1]

$$\sigma_{fw1} = \sigma_{w1} - 2\mu p_0[1 - \exp(-S/k)] \qquad (9.1)$$

where σ_{w1} is the fatigue strength in the absence of fretting, μ is the coefficient of friction, p_0 is the contact pressure, S is the amplitude of slip and k is a constant. The term $2\mu p_0[1 - \exp(-S/k)]$ represents the additional stress in the surface due to the fretting contact and is greater the higher the coefficient of friction and the pressure, as would be expected. The exponential term becomes insignificant at larger values of the slip amplitude: with $\mu = 0{\cdot}67$, $p_0 = 500\ \mathrm{MN\,m^{-2}}$, $\sigma_{w1} = 736\ \mathrm{MN\,m^{-2}}$ and $k = 3{\cdot}38 \times 10^{-3}$, σ_{fw1} reaches a constant value of $90\ \mathrm{MN\,m^{-2}}$ when the slip amplitude is greater than 20 μm. This is in reasonably good agreement with the experimental finding of Field and Waters that the critical value of the slip amplitude at which the fatigue strength reached its minimum value for steel specimens was between 8 and 14 μm.[2] In a recent paper Duquette reported that at a given alternating stress the fatigue life is a minimum at an amplitude of slip of 20 μm over a range of clamping pressures from 20 to 41 MN m^{-2}.[3] At higher amplitudes of slip the fatigue strength increases again due to the excessive wear. This has been attributed to a reduction in coefficient of friction by the presence of wear debris and also to the surface layer in which a fatigue crack is initiated being continually abraded away before the crack has time to propagate.

Nishioka and Hirakawa found a linear relationship between pressure and the alternating stress to initiate a fatigue crack over a range of pressure 0–400 MN m^{-2}, although the effect of pressure in reducing the fatigue strength was only apparent at pressures up to 180 MN m^{-2} (Fig. 9.1).[4] This is in agreement with the graph shown in an earlier monograph which shows that increasing the contact pressure has no further effect above a value between 70 and 150 MN m^{-2} depending on the nature of the material.[5]

The main drawback to the Nishioka and Hirakawa expression is that it implies that if the fatigue strength σ_{w1} is increased there will be a

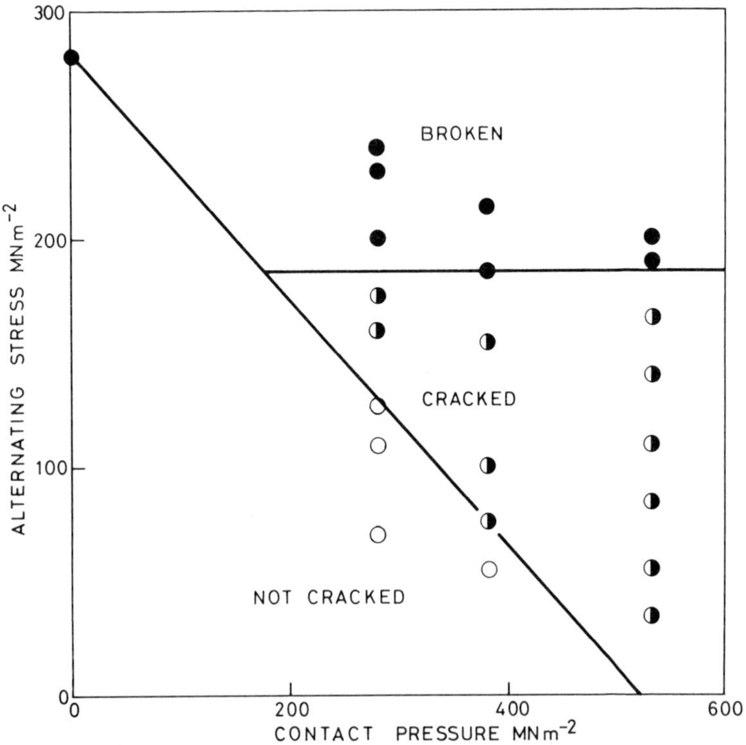

FIG. 9.1. Effect of clamping pressure on initiation of fretting fatigue cracks.[4]

corresponding increase in the fretting fatigue strength. Table 10·1 (Chapter 10) demonstrates clearly that this is not the case because fretting is analogous to the effect of a notch and materials of high strength are generally more notch sensitive. This is taken into account in the fracture mechanics approach of Dr Edwards which forms the subject of Chapter 3. The nature and magnitude of stresses within a fretting contact have also been investigated in a more sophisticated way with finite element analysis as described in Chapter 2 and other publications.[6,7,8] Nevertheless, the simpler analysis of Nishioka and Hirakawa has been shown in their numerous publications to be in good agreement with the observed generation of propagating and non-propagating cracks.

9.2 DIRECT OBSERVATION OF FRETTING DAMAGE

The scanning electron microscope has proved a valuable tool in examining the initial stages and the development to final failure of fatigue damage promoted by fretting. Most recent interpretations of crack initiation and general surface damage are based on such observations. However,

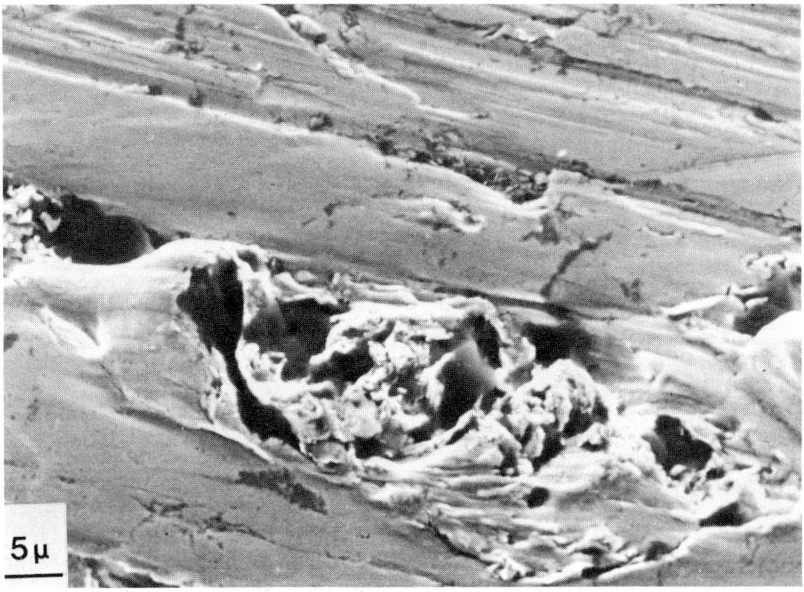

FIG. 9.2. Areas of local welding on fretted surface of 0·2C steel after 500 cycles of fretting.

metallographic sections through the surface can still provide considerable evidence of events in the surface layers. The technique which combines both methods of analysis at once, namely the taper section developed by Eyre, enables sub-surface features to be related to surface damage.[8]

Observations of the damage on mild steel indicate that in the early stages local welds occur causing considerable roughening of the surface (Fig. 9.2). This is followed by smearing of material and a reduction in the surface roughness (Fig. 9.3). Eventually material is removed from the surface by a

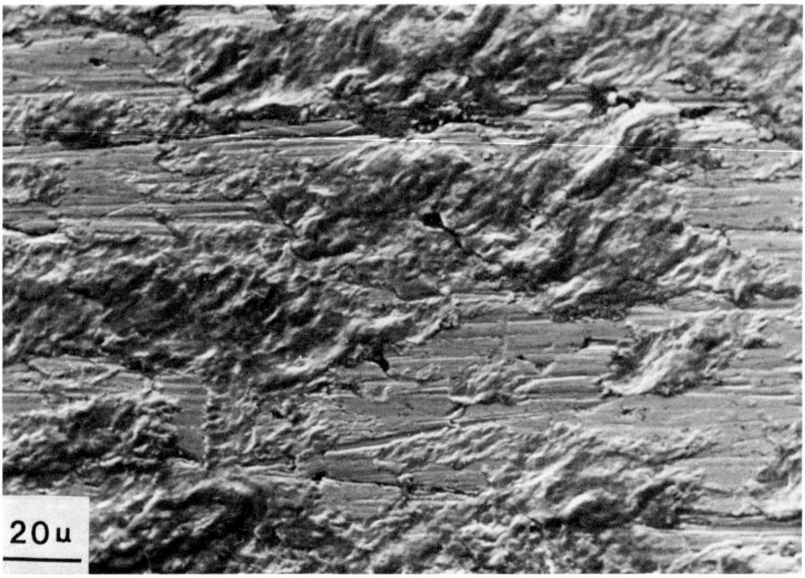

FIG. 9.3. Smeared material on fretted surface of 0·2C steel after 5000 cycles of fretting.

delamination process producing plate-like particles of debris (Fig. 9.4). At this stage the debris consists of thin oxide-covered metal particles. They are then ground between the two surfaces to form finer particles with a consequent increase in the oxide content.[9]

At what stage in this sequence of events is the first fatigue crack initiated, and at what rate does it grow? Acoustic emission studies on an aluminium alloy show that small cracks are initiated by fretting very early in the fatigue life, i.e. about 0·1 % of the total life, or a few thousand cycles based on a run-out life of 10^7 cycles.[10] After initiation there was a period of very slow propagation in which the ΔK v. da/dn plot was linear with a crack growth rate of $1{\cdot}3 \times 10^{-8}$ mm/cycle at stresses near the fretting fatigue limit.

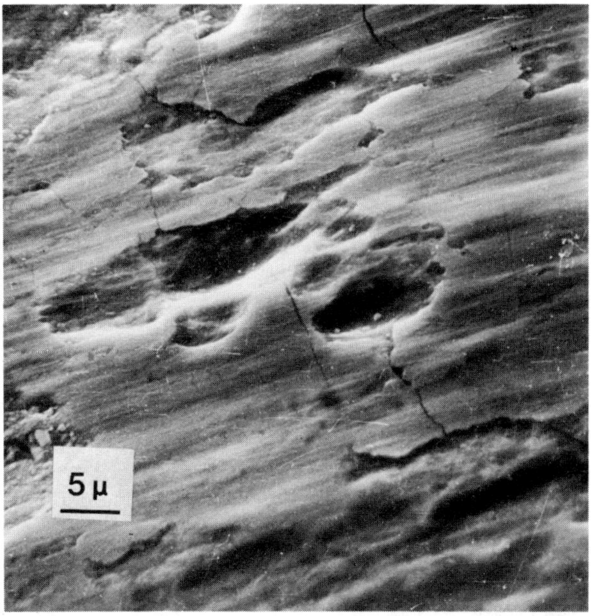

FIG. 9.4. Delamination on fretted surface of 0·7C steel after 10^5 cycles of fretting.

Finally the crack propagated according to the Paris–Erdogan power law. Alic and Hawley,[11] measuring crack growth rates from examination of striations on the fretting fatigue fracture surface, found that the crack growth was accelerated in the initial 100 μm of its growth, but suggest that this could be partially the effect described by Pearson[12] and by Chauhan and Roberts[13] of the crack growth rate of very short cracks being considerably greater than that of long cracks.

The author has determined the number of fretting cycles to initiate a propagating fatigue in various materials and has generally found that this is about 10 % of the fatigue life, but this period apparently includes the stage observed by Ryman and Blackwell when the crack is growing very slowly. This initiation stage coincides with the development of maximum adhesion between the surfaces in some materials, e.g. mild steel and austenitic stainless steel (Fig. 9.5).

The site of the initial fatigue crack where slip is occurring over only part of the surface area in contact, i.e. the situation of micro-slip, has been observed to be in the boundary between the region of slip and non-slip (Fig. 9.6). Development of debris can cause this boundary to move and is

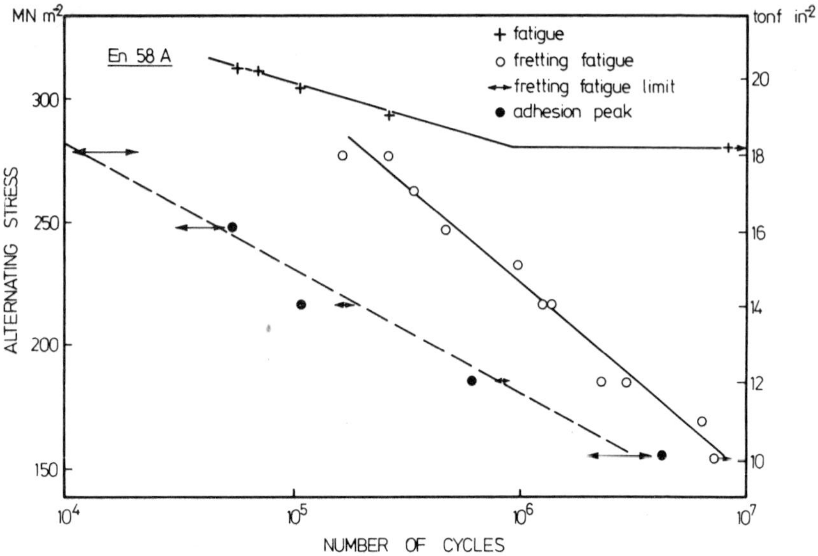

FIG. 9.5. The relation between the number of cycles of fretting to initiate a propagating fatigue crack, the number of cycles to promote maximum adhesion between the surfaces, and the total life in fretting fatigue for an annealed austenitic stainless steel.

FIG. 9.6. Fatigue crack in boundary between slip and non-slip regions on fretted surface of 0·7C steel.

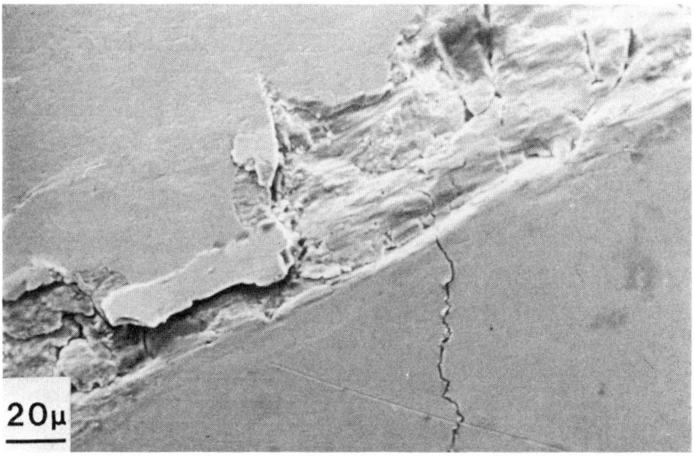

FIG. 9.7. Delamination and associated fatigue cracks on fretted surface of Ti–6Al–4V, produced by tangentially vibrating a heavily loaded (7500 N) hardened steel ball (diam. 19 mm) in contact with the surface after 3×10^5 cycles of fretting.

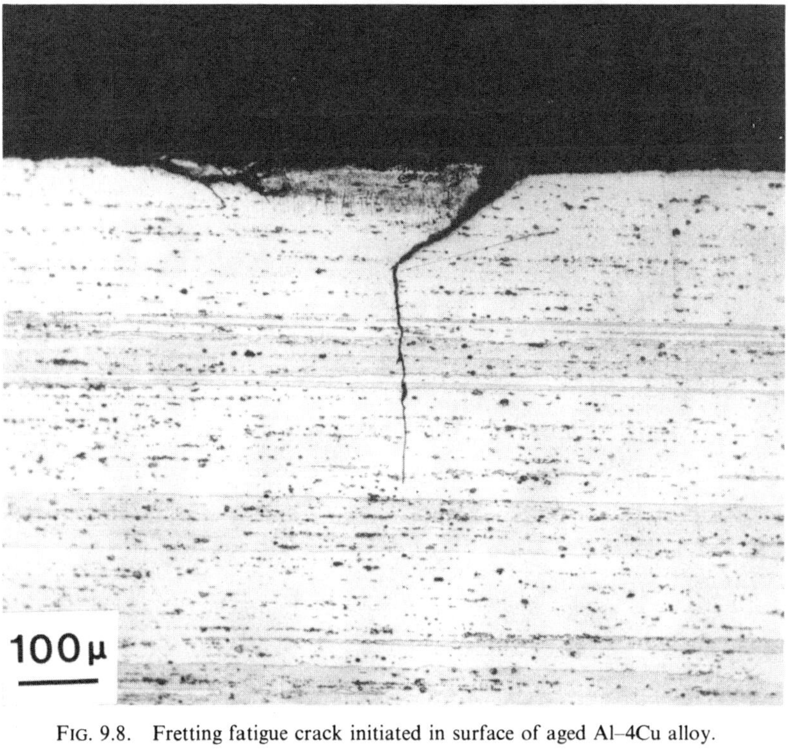

FIG. 9.8. Fretting fatigue crack initiated in surface of aged Al–4Cu alloy.

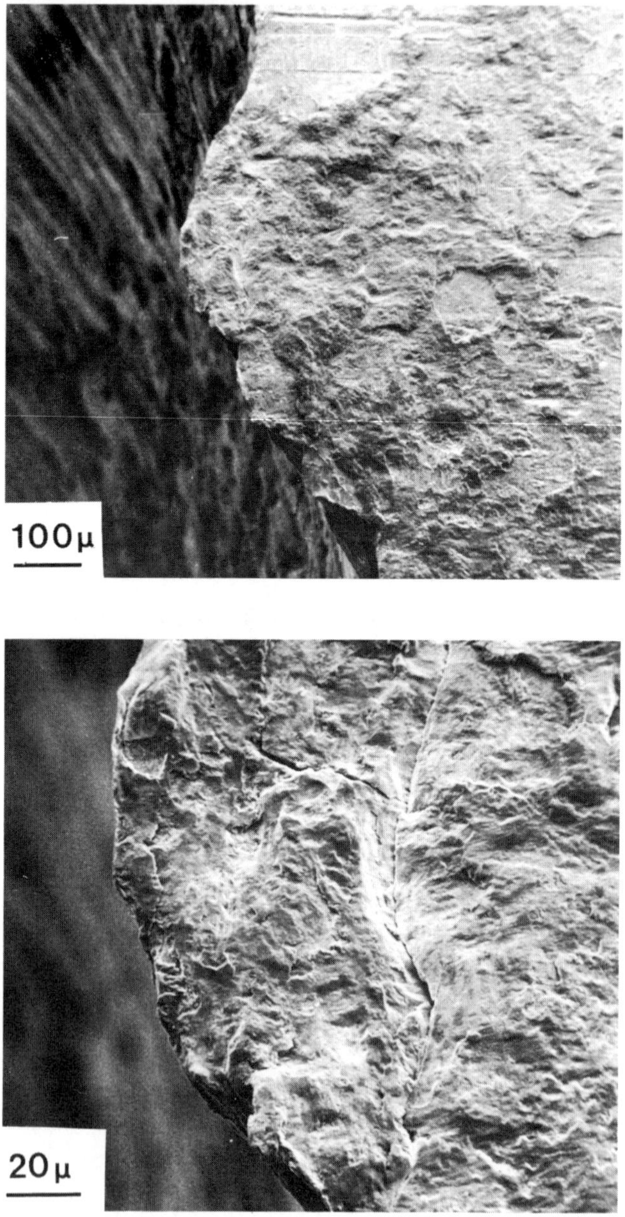

(a)

(b)

FIG. 9.9. (a) Site of fretting fatigue failure on a press-fitted steel shaft showing typical tongue, (b) at higher magnification showing subsidiary cracks.

thought to be the reason for the generation of multiple cracks which is a feature of fretting fatigue failures. Where the contact area is limited there is a stress concentration at the end of the region of contact and cracks are initiated just within the contact area near the boundary of the fretted region.[14] There is evidence from photoelastic investigations that the slip/non-slip boundary does move, particularly if the tangential load is increased, which is a likely occurrence where the fatigue loading is random rather than of constant amplitude.[15]

There has been some discussion whether the sub-surface cracks which are a feature of the delamination process could change direction and become a propagating fatigue crack into the bulk of the material. The author was of the opinion that the fatigue cracks which cause failure had been initiated before the delamination process had progressed to any extent. Alic and Hawley make a clear distinction between the two types of crack.[11] However, in recent work involving a very heavily loaded contact consisting of a hardened steel ball pressed into flat specimens of titanium alloys and vibrated tangentially, fatigue cracks have been seen growing from areas of delamination (Fig. 9.7).[16] The susceptibility of three titanium alloys to the development of such cracks is described in Section 9.3.

Cracks, once initiated, follow a course at an angle to the surface, but when the crack is beyond the influence of the fretting action it changes direction becoming perpendicular to the principal alternating stress (Fig. 9.8). This figure illustrates another feature of fretting fatigue cracks in that the initial oblique portion of the crack is opened due to the crack relieving the stress concentration at this point in the surface. It frequently fills up with fretting debris which contributes to the extent to which it is forced open. On the final fracture surface the oblique portion of the crack appears as a tongue on one surface and a matching chip on the other. Figure 9.9(a) and (b) show such a tongue on a fatigue fracture of a steel shaft carrying a press-fitted collar. In addition to the main crack which resulted in failure numerous other cracks are visible (Fig. 9.9(b)).

9.3 THE ROLE OF MICROSTRUCTURE

It has been known for some time that cast structures are less susceptible to fretting fatigue than forged structures.[17] Also that annealed material is less susceptible than work-hardened.[18] The effect of microstructure in multi-phase alloys has not been extensively investigated, although it has been studied in relation to normal fatigue failure and crack propagation.

Considerable energy is dissipated in the fretting contact. Indications of plastic deformation are frequently seen and it has been suggested that there are local temperature rises which may amount to several hundred degrees Centigrade.[19] Transmission electron microscope observations on the fretted surface of an austenitic stainless steel reveal that mechanical twinning and the formation of the hexagonal phase ε-martensite have occurred.[20] The author has also seen evidence of twinning in the α phase of the titanium alloy Ti–6Al–4V produced by fretting.[21] Whether or not there is a temperature rise is controversial. Recent careful work by Sproles and Duquette on the fretting of mild steel shows that the measurable temperature rise is only of the order of 20°.[22] However, structural changes do occur in the surface layers which in some cases are similar to the effects produced by heating, e.g. softening attributed to over-ageing has been observed in an aged Al–4Cu–1Mg alloy.[23] It is important, therefore, that an alloy in a heat-treated and metastable condition should remain stable under the fretting action.

Reeves and Hoeppner have compared the fretting fatigue behaviour of a carbon steel in the normalised (ferrite + pearlite) condition and the martensitic condition.[24] They conclude that the latter structure is less affected by fretting fatigue due to its greater hardness and wear resistance. The three titanium alloys referred to in Section 9.2 are all α + β alloys but with somewhat different microstructures. Their response to fretting under a heavy load as judged by the number of propagating fatigue cracks emerging

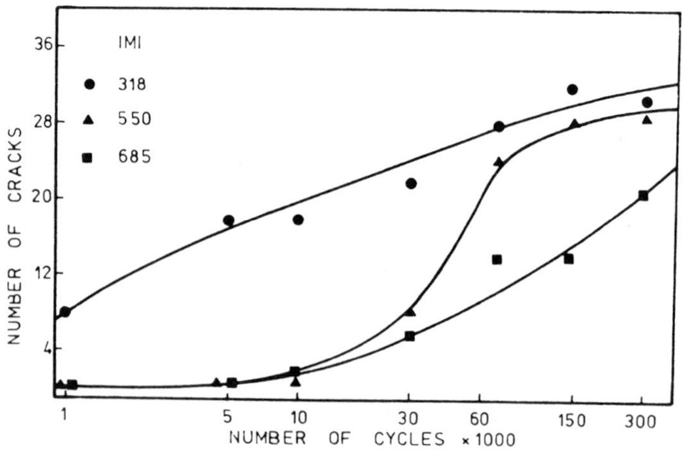

FIG. 9.10. Initiation of fatigue cracks in three titanium alloys; IMI 318 Ti–6Al–4V, IMI 550 Ti–4Al–4Mo–2Sn–0·5Si, IMI 685 Ti–6Al–5Zr–0·5Mo–0·25Si.

from the fretted area is shown in Fig. 9.10. The details of the heat treatments of the three alloys are as follows:

Ti–6Al–4V—annealed 700 °C 2 h, air cooled;
Ti–4Al–4Mo–2Sn–0·5Si—solution heat treated 900 °C 0·5 h, air cooled, aged 500 °C 24 h, air cooled, aged 500 °C 24 h, air cooled;
Ti–6Al–5Zr–0·5Mo–0·25Si—solution heat treated 1050 °C 0·5 h, oil quenched, aged 500 °C 24 h, air cooled.

The resultant microstructures are shown in Fig. 9.11. It is evident that the microstructures are progressively finer, the third alloy consisting of an acicular structure. The mechanical properties of the alloys are shown in Table 9.1 and their behaviour in fretting fatigue tests in Table 9.2. Their properties, apart from impact strength and fatigue strength, are similar As is often the case, the material with the lower fatigue strength is less influenced by fretting, although the low impact strength suggests a low fracture toughness which would lead one to expect a heightened susceptibility to fretting fatigue failure.

TABLE 9.1

MECHANICAL PROPERTIES OF THE THREE TITANIUM ALLOYS, IMI 318 (Ti–6Al–4V), IMI 550 (Ti–4Al–4Mo–2Sn–0·5Si) AND IMI 685 (Ti–6Al–5Zr–0·5Mo–0·25Si)

	318	550	685
UTS ($MN\,m^{-2}$)	980	1 124	1 057
0·2% PS ($MN\,m^{-2}$)	911	1 023	926
elongation (%)	18	16	11
reduction in area (%)	44	53	25
hardness (VHN)	365	361	369
impact strength (Nm)	10·4	22·5	6·0
fatigue strength at 10^7 cycles ($MN\,m^{-2}$)	539	647	417

TABLE 9.2

BEHAVIOUR IN FRETTING FATIGUE TESTS OF THREE TITANIUM ALLOYS

	318	550	685
Reduction in fatigue strength due to fretting (at 10^7 cycles) (%)	50	62·5	35
Number of fretting cycles to initiate a propagating fatigue crack at an alternating stress of ± 278 $MN\,m^{-2}$	50 000	75 000	75 000
Proportion of total life spent in crack initiation at an alternating stress of ± 278 $MN\,m^{-2}$ (%)	16	27	37

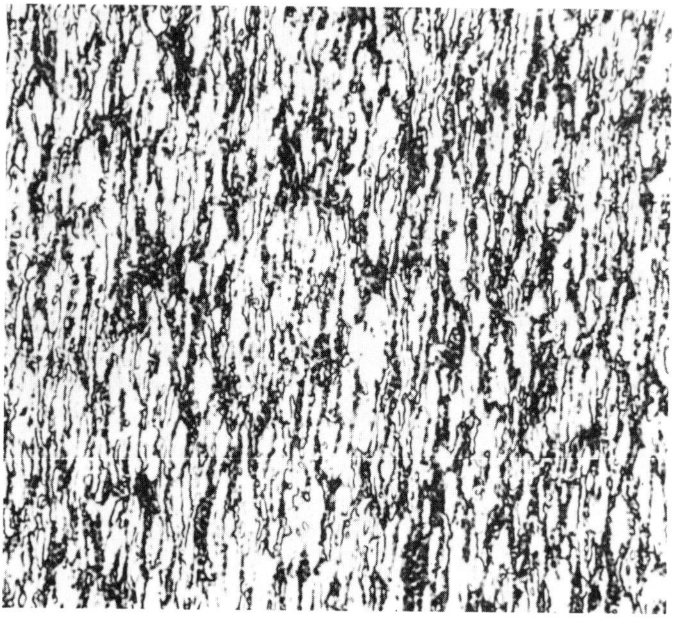
(a)

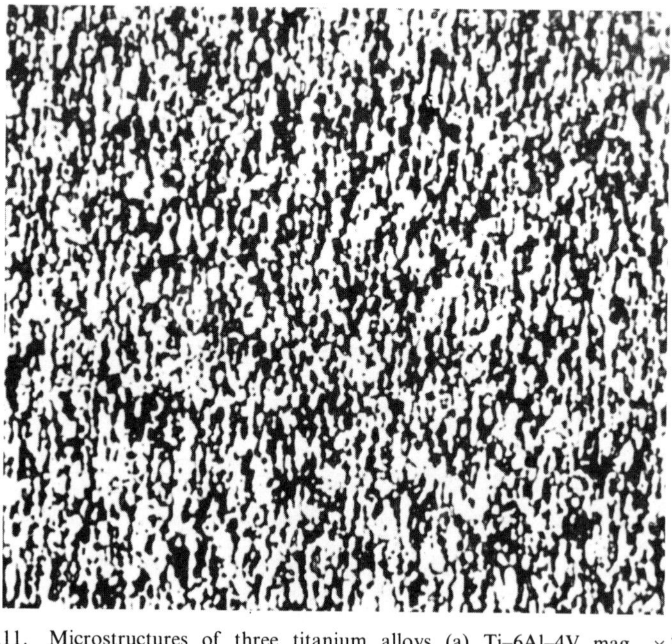
(b)

FIG. 9.11. Microstructures of three titanium alloys (a) Ti–6Al–4V mag. ×325 (b) Ti–4Al–4Mo–2Sn–0·5Si mag. ×325 (c) Ti–6Al–5Zr–0·5Mo–0·25Si mag. ×325.

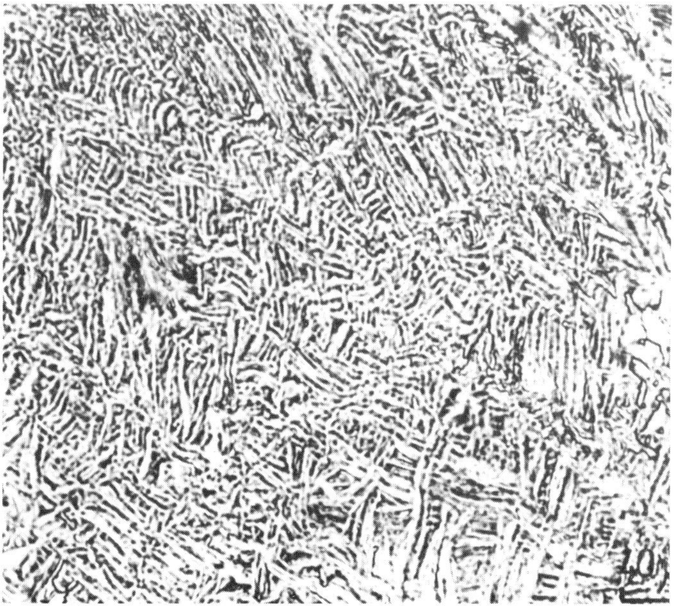

(c)

FIG. 9.11—*contd.*

It is obvious that the effect of microstructure on fretting fatigue, particularly in titanium alloys which are particularly prone to this type of damage, has been insufficiently investigated. A necessary adjunct to such studies is the development of methods to prepare thin sections from fretted regions for examination in the transmission electron microscope. This is being undertaken in the author's laboratory and some early results have already been reported.[21] The purpose of the work is to study dislocation configurations, phase changes, and signs of crack nucleation in the surface layer with the intention of developing structures and surface treatments which will be resistant to crack initiation.

9.4 MODE OF STRESSING

One feature of fretting fatigue which throws some light on the processes occurring is that the reduction in fatigue strength is lowest when the fatigue testing is carried out in torsion, and greatest when the testing is in rotating–bending or plane-reverse–bending.[25,26] The reduction when testing in push–pull is intermediate between these two extremes.[6] Bramhall has provided an explanation of these observations. Figure 9.12 represents in much exaggerated form the state of affairs at the contact region between the bridge foot and the specimen surface in push–pull and in rotating–bending. Lifting away occurs in both cases but is much more pronounced in rotating–bending. In the latter case most of the fretting action occurs when the specimen surface is in compression. Earlier Collins and Marco had shown in tests where specimens were first subjected to fretting damage and then fatigued, that the reduction in fatigue strength

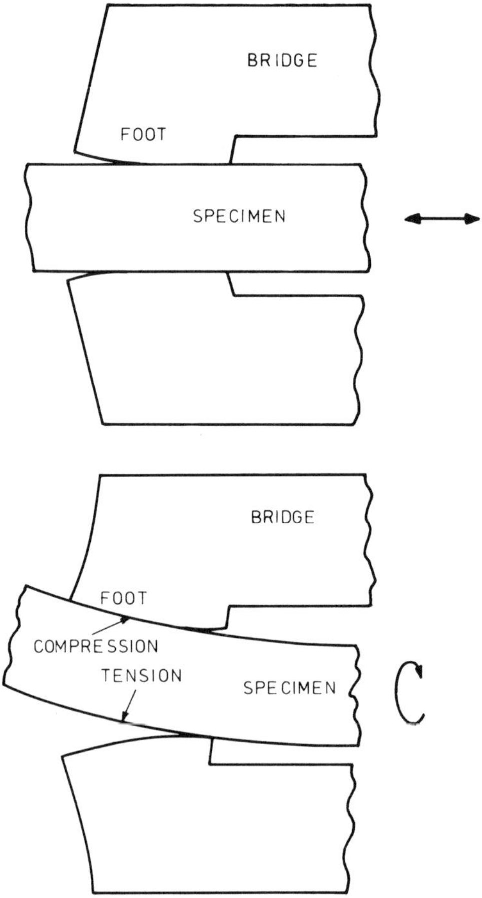

FIG. 9.12. Schematic representation, much exaggerated, of the contact bridge foot and fatigue specimen: top, in push–pull at peak tensile load; bottom, in rotating–bending.

was much greater when a mean compressive load was applied in the fretting part of the test than when a tensile mean load was applied.[27] Their explanation of this somewhat unexpected effect was that the plastic deformation caused by the fretting action led to a local relaxation of the imposed mean stress. On removing the mean stress the fretted region was left with an internal stress of opposite sign to the mean stress. Where the mean stress was compressive this would result in a local tensile stress and would account for the large reduction in fatigue strength. This effect would be much more pronounced in rotating–bending or plane-reverse–bending than in push–pull. The same argument would apply to a press-fit on a shaft operating under rotating–bending or push–pull. In torsion, on the other hand, there would be no lifting away and although there is an alternating surface compressive stress in this mode of testing it is negligible compared with the compressive stress in the other types of test. The detrimental effect of a compressive stress is of course only apparent where that stress is periodic. Induced surface compressive stresses, produced by shot peening or nitriding which are in-built and remain unchanged are beneficial in countering fretting fatigue as is described in Chapter 10.

The question whether the fretting action can induce residual stresses is answered by Professor Endo in Chapter 5, but the other aspect, i.e. whether fretting can eliminate existing residual stresses, has not been investigated except for the circumstantial evidence provided by Collins and Marco. To obtain more quantitative data an investigation is in progress in the author's laboratory to determine by the X-ray back-reflection method what effect fretting has on residual stresses produced by various mechanical treatments on the surface of an age-hardened aluminium alloy.

9.5 CONCLUSION

The main factor in producing a fretting fatigue failure is the additional alternating shear stress in the contact areas arising from the frictional stress between the surfaces. The initiation sites for fatigue cracks are points where there is a particular stress concentration, e.g. at the boundary between slip and non-slip regions or at or near the outer contact boundary. The pitting and roughening produced by local cold welding does not appear to play a significant role in crack initiation because this occurs in fretting *in vacuo* but cracks are not initiated as easily as when the environment is air. The importance of the environment in promoting fretting fatigue failure is stressed by Poon and Hoeppner who find that cracks are initiated from corrosion pits in an aluminium alloy.[28] However, the author is of the opinion that pitting is not a major factor, since surface roughening can be produced by fretting an aluminium alloy against an abrasive material such as fused alumina without affecting the fatigue properties.[29]

The initial rate of growth of the fatigue crack is increased by the additional alternating shear stress but it slows down when it propagates out of the region influenced by the fretting contact. Signs of severe high-strain fatigue are often visible on the surface with evidence of plastic deformation. On some ductile materials lines of holes develop and curious related phenomena are seen such as the formation of the 'zip fastener'.

Microstructure has an influence on fretting fatigue. In titanium alloys a fine or acicular $\alpha + \beta$ structure is more resistant to damage than a coarser annealed structure. The mode of stressing also has an influence, torsion being least damaging, followed by axial loading (push–pull), with rotating–bending or plane-reverse–bending having the maximum effect.

REFERENCES

1. Nishioka, K. and Hirakawa, K. *Bull. J.S.M.E.*, **12**, 692–7 (1969).
2. Field, J. E. and Waters, D. M. N.E.L. East Kilbride, Report No. 120 (1963).
3. Duquette, D. J. *Strength of metals and alloys*, Pergamon, Oxford, p. 214 (1980).
4. Nishioka, K. and Hirakawa, K. *Bull. J.S.M.E.*, **15**, 135–42 (1972).
5. Waterhouse, R. B. *Fretting corrosion*, Pergamon, Oxford, p. 139 (1972).
6. Wright, G. P. and O'Connor, J. J. *Proc. Instn. Mech. Engrs.*, **186**, 827–35 (1972).
7. Doeser, B. A. and Waterhouse, R. B. *Proc. Euromech. Colloquium No. 110*, Linköping Institute of Technology, Sweden, September 1978.
8. Eyre, T. S. and Dutta, K. *Lubricn. Engng.*, **31**, 521–9 (1975).
9. Waterhouse, R. B. *Wear*, **45**, 355–64 (1977).
10. Ryman, R. J. and Blackwell, M. P. An Acoustic Emission Investigation of Fretting Fatigue Cracks in BSL 65 Aluminium Alloy under Constant and Random Amplitude Loading, Report No. FAT/130, Sept. 1977, Testwell Ltd, Daventry.
11. Alic, J. A. and Hawley, A. L. *Wear*, **56**, 377–89 (1979).
12. Pearson, S. *Engng. Frac. Mech.*, **7**, 235–47 (1975).
13. Chauhan, P. and Roberts, B. W. *The Metallurgist and Mat. Technologist*, **11**, 131–6 (1979).
14. Alic, J. A., Hawley, A. L. and Urey, J. M. *Wear*, **56**, 351–61 (1979).
15. Rollins, C. T. and Sandorff, P. E. *J. of Aircraft*, **9**, 581–4 (1972).
16. Goto, S. and Waterhouse, R. B. *Proc. 4th Internat. Conf. on Titanium*, Kyoto, Japan, May 1980.
17. Sachs, G. and Stefan, P. *Trans. A.S.M.*, **29**, 373–401 (1941).
18. Wharton, M. H., Taylor, D. E. and Waterhouse, R. B. *Wear*, **23**, 251–60 (1973).

19. Alyab'ev, A. Ya., Kazimirchik, Yu. A. and Onoprienko, V. *Sov. Mater. Sci.*, **6**, 284–6 (1970).
20. Alyab'ev, A. Ya., Shevelya, V. V. and Rozhkov, M. N. *Sov. Mater. Sci.*, **6**, 672–5 (1970).
21. Waterhouse, R. B. *Fundamentals of tribology*, ed. N. P. Suli and N. Saka, MIT Press, Cambridge, Mass. pp. 567–84 (1980).
22. Sproles, E. S. and Duquette, D. J. *Wear*, **47**, 387–96 (1978).
23. Waterhouse, R. B. *Fretting in Aircraft Systems*, AGARD Conf. Proc. No. 161, 1975 8.1-8.
24. Reeves, R. K. and Hoeppner, D. W. *Wear*, **47**, 221–9 (1978).
25. Endo, K., Goto, H. and Nakamura, T. *Bull. Jap. Soc. Mech. Engrs.*, **12**, 1300–8 (1969).
26. Bramhall, R. *Studies in Fretting Fatigue*, Ph.D. Thesis, Oxford University, November 1973.
27. Collins, J. A. and Marco, S. M. *Proc. ASTM*, **64**, 547–60 (1964).
28. Poon, C. and Hoeppner, D. W. *Wear*, **52**, 175–91 (1979).
29. Waterhouse, R. B. *Wear*, **58**, 391–2 (1980).

10

Avoidance of Fretting Fatigue Failures

R. B. Waterhouse

Department of Metallurgy and Materials Science,
University of Nottingham, UK

10.1 INTRODUCTION

In any situation involving fatigue the general principles of minimising fatigue failures apply. Of prime importance is the avoidance of stress raisers. This is achieved in practice by close attention to design, by insisting on a high standard of surface finish, and by taking anti-corrosion measures to prevent possible pitting corrosion. Other measures, such as putting the surface layers into compression by mechanical means, e.g. surface rolling or peening, or by chemical means, e.g. nitriding, are also to be generally recommended. The additional treatments which may be applied in the case of fretting fatigue arise from the damage which initiates the fatigue crack occurring at contacting surfaces. This damage, as seen in the previous chapter, arises from the transfer of shear stresses across the interface which are alternating since they arise from the cyclic stressing responsible for the fatigue situation. Where these shear stresses exceed the constraints of friction, slip occurs and fretting ensues. Since the magnitude of the shear stresses is very much determined by the relation of the contact region to regions of stress concentration, design is necessarily an important factor in minimising fretting fatigue failures. However, the fact that the surfaces in the contact are damaged by the process leads to the introduction of a wide variety of surface treatments to overcome this feature of fretting fatigue, most of which would not be applied in the case of simple fatigue, and some of which actually have a deleterious effect on the fatigue properties.

The methods which have been applied up to the year 1970 were reviewed extensively in a previous publication,[1] and these will only be briefly referred to in order to indicate the principles of the process. The major part of this chapter is concerned with developments in the last ten years. However, in this general introduction reference should be made to two recent review articles on the subject.[2,3]

10.2 DESIGN

Obviously one way of avoiding fretting problems would be to do away with contacting surfaces as far as possible, e.g. by replacing bolted flange connections by a single machined member or by welding, although welds themselves can create problems where fatigue is concerned. However, the restrictions placed on the use of single-unit components by the need for ease of assembly and dismantling of the whole installation usually limit the possibilities in this direction. It should be pointed out that single-unit construction can be less attractive economically since if one part of the unit fails by fatigue or surface wear, then the whole must be replaced, whereas in an assembly of smaller components only the affected part needs replacing. These are factors which need to be considered by the designer.

A possible half-way house is the use of certain epoxy resins which are sold commercially for applying to nuts and bolts to prevent their loosening in service. They have evocative names which confidently proclaim their function. The author has seen examples where this method has been applied successfully, e.g. where fretting occurred between a bearing housing and its support. The disadvantage in their use is if the assembly has to be dismantled, when the efficacy of these products is manifestly demonstrated.

In general, contacting surfaces cannot be avoided and the guiding principle in avoiding fretting fatigue problems should be to reduce stress

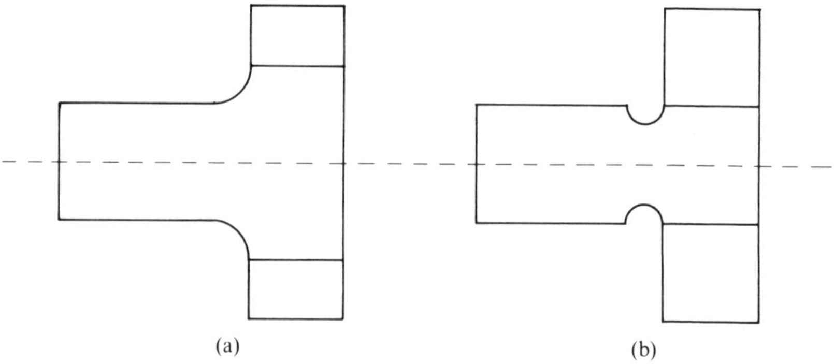

FIG. 10.1. (a) Preferred design of a wheel-on-axle assembly to minimise stress concentration at the interface; (b) Stress-relieving groove in wheel-on-axle assembly.

concentrations in the vicinity of the contact. The classic example is that of the wheel-on-axle assembly where the diameter of the wheel seat should be greater than that of the axle with a generous fillet radius between the two (Fig. 10.1(a)). If this is not possible then a stress relieving groove adjacent to the hub achieves the same objective (Fig. 10.1(b)).

A common situation where fretting leads to failure under fatigue conditions is the pin-loaded joint (Fig. 10.2). When such a joint is tested in

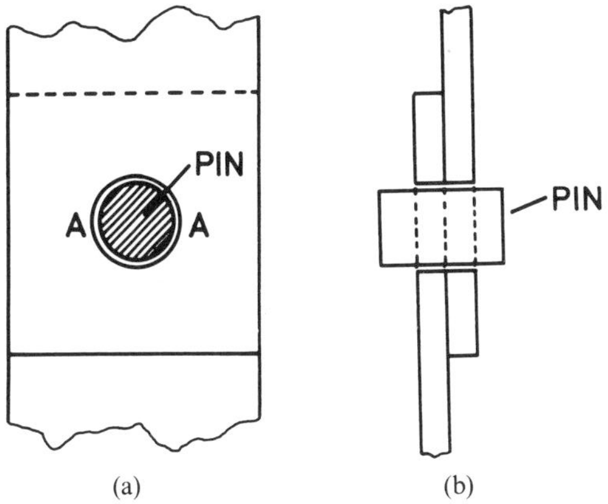

FIG. 10.2. (a) Front elevation of single pinned joint; (b) Side elevation of single pinned joint.

fluctuating tension, fretting occurs at the diametrically opposite sides of the hole, A–A, and a fatigue crack is initiated which propagates through the reduced section to failure. By machining flats on the sides of the pin, contact can be avoided at this critical area. The magnitude of the flat is indicated by the angle α in Fig. 10.3. The effect of the flat has been investigated by White and is shown in Fig. 10.4.[4] Under pulsating tension ($R = 0$) the fatigue strength increases linearly as α is increased to 40°. At this value of α the

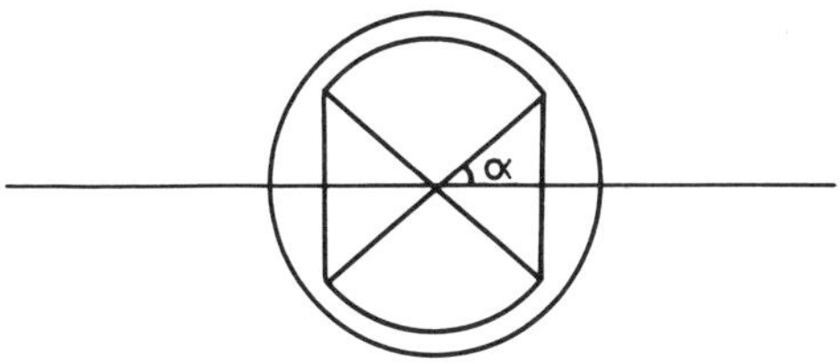

FIG. 10.3. Flats machined on sides of the pin to prevent contact at the position of maximum stress. The size of flat depends on angle α.

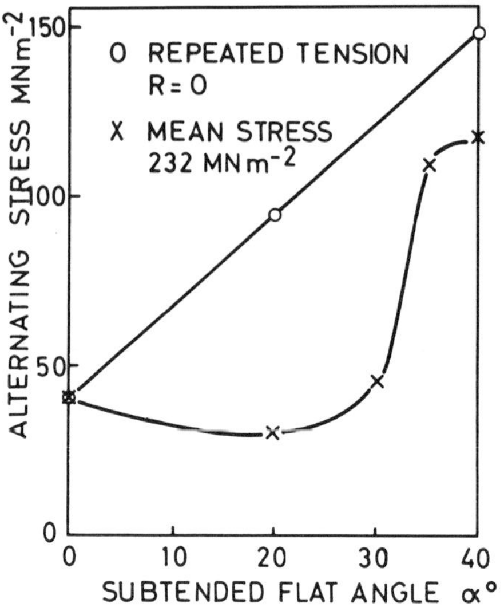

FIG. 10.4. Relation between fatigue strength of a pinned joint and the angle α for specimens tested in fluctuating tension and with a mean stress of 232 MN m^{-2}.[4]

fatigue failures were occurring in unfretted material and, therefore, any further increase in α would not give any further improvement. Where a mean tensile stress was applied no benefit accrued from the flat until the value of α was 35°.

Another possible variable in the pinned joint is the degree of interference fit of the pin in the hole. Buch has investigated this effect by comparing the stress concentration factor K_f for specimens with an open hole and a pinned joint with the same size hole.[5] At fatigue lives greater than 10^6 cycles, K_f for the pinned joint was 2·90 compared with 2·65, the difference being attributed to the effect of fretting. At lives less than 10^5 cycles the strength reduction factors were identical for the joint and the open-hole specimen. This confirms the view that fretting has a much greater effect on fatigue strength at low stresses and long lives, i.e. high cycle fatigue.[6] The stresses arising from the interference fit result in a reduced stress amplitude but contribute to the higher mean stress. At a critical value of the nominal tensile stress these two effects are equal and therefore any beneficial effect is cancelled out. This critical value is given by

$$\sigma_K \geq \frac{2\sigma_i}{K_t} \tag{10.1}$$

where σ_i is the internal stress due to the interference fit and is taken as the 0·2% proof stress, and K_t is the theoretical stress concentration factor for the hole. The critical value of the tensile stress is raised with a higher interference stress and a lower stress concentration factor.

The purpose of pinned and bolted joints is to transfer the load from one member to another by ensuring the distribution of the resultant shear load over as great an area of the contacting surfaces as possible. This demands a high frictional force which depends on the clamping pressure and the coefficient of friction. This means that the methods of reducing fretting damage by encouraging a low coefficient of friction between the surfaces or incorporating inserts of material with low elastic modules are excluded. This is evident in the behaviour of double shear bolted lap joints with friction-reducing inserts of polytetrafluoroethylene and nylon containing molybdenum disulphide.[7] Fretting was prevented at the contacting surfaces but the fatigue strength of these joints was the same as that of the untreated joints because the bulk of the load was transferred to the pins.

A form of design similar to the pinned joint is the bonded double strap joint. In this case the two members to be jointed have strips fixed to the join by a suitable adhesive. If the ends of the straps are square there exist stress concentrations at these points (Fig. 10.5(a)). These may be minimised by bevelling the ends of the strap (Fig. 10.5(b)). This modification in design has been investigated in an aluminium alloy using a low temperature curing adhesive cured at about 115 °C.[8] The joints were tested in static loading as well as fatigue loading. In static tests more failures occurred in the adhesive

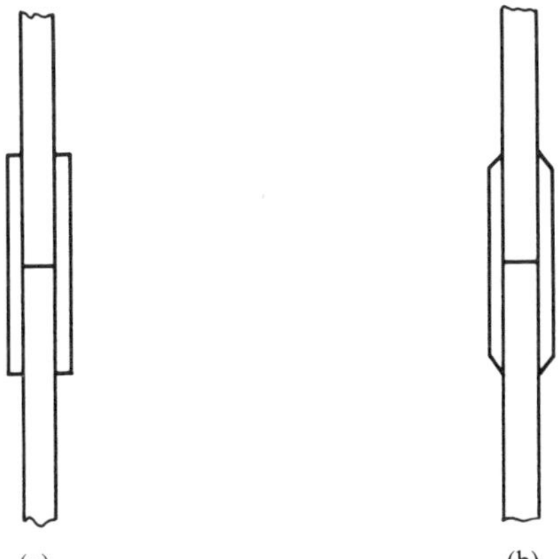

(a) (b)

FIG. 10.5. Design of a bonded double strap joint (a) square ended straps, (b) bevelled straps to reduce the stress concentration at the ends of the straps.

with the bevelled straps, but in the fatigue tests adhesive failure was predominant at high alternating stresses where the bevelled joints were slightly superior to the square-ended. At lower stresses more failures occurred in the metal with the bevelled joint, indicating that the stress concentration had been reduced. Fifty percent of the failures occurred at the strap edge in the square-ended joints compared with only 23% in the bevelled joints.

Another type of design which often gives rise to fretting fatigue failures is the dovetail fixing which is very common in axial compressors as a means of attaching the rotor blades to the disc (Fig. 10.6).[9]

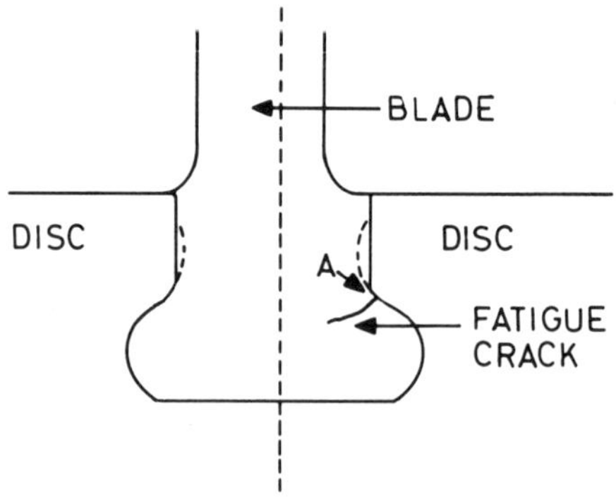

FIG. 10.6. Dovetail fixing of a rotor blade to the disc in an axial compressor. Fretting initiates a fatigue crack at point A. This is minimised by undercutting the neck (shown dotted).

Fretting at the upper edge of the bearing surface A initiates a small crack which may propagate to failure if the alternating bending stress due to blade vibration is large enough. Undercutting of the neck of the dovetail indicated by the dotted lines increased the fatigue strength of the component by a factor of two. This type of fixing is often used in securing the poles to the rotors in electrical generators. The dovetail is tightened with flat tapered keys on the sloping surfaces which pull the pole shoulders down on to the rotor, putting the dovetail 'neck' in tension (Fig. 10.7). Fretting at the edges of the keys, point A, can result in initiation of fatigue cracks.[10] In this case also the provision of a relief radius at the key edge greatly improved the fatigue behaviour (Fig. 10.8). Figure 10.9 shows the fatigue curves for the two types of design. It was noted that in the original design, failure by fretting fatigue occurred at lives greater than 10^6 cycles, but at shorter lives failure was by normal fatigue in the dovetail neck.

A more general aspect of the influence of design is revealed in considering

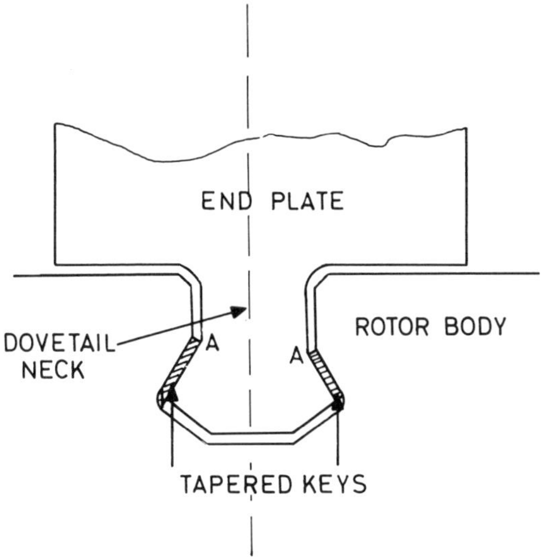

FIG. 10.7. Dovetail fixing of a pole piece end plate to the rotor in an electrical generator. Fretting initiates cracks at point A.[10]

the actual geometry of the contacting surfaces. The author had shown that in a contact consisting of a cylindrical surface in contact with a plane, where the maximum pressure was kept constant, the broader the contact, the greater the reduction in fatigue life due to fretting.[1] This was attributed to the greater volume of material being influenced by the tangential stresses in the broader contact. This principle has been applied more generally to

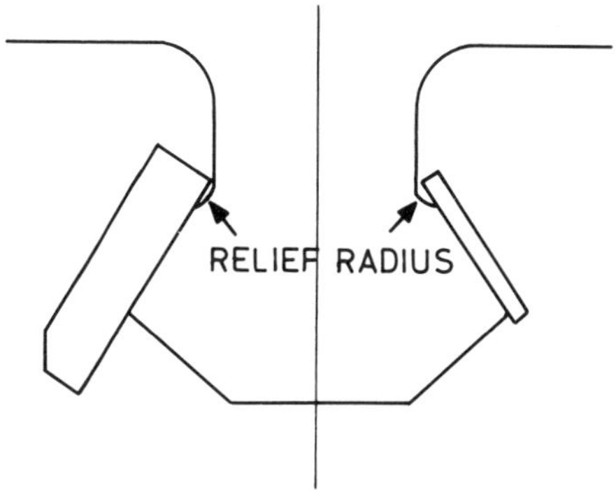

FIG. 10.8. Modified design—provision of relief radius to reduce the stress concentration at the ends of the tapered keys.

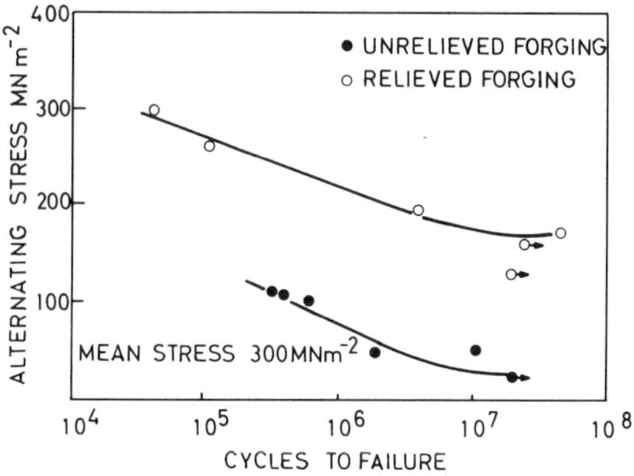

FIG. 10.9. Fatigue curves for dovetail-disc fixing in forged steel with and without a relief radius.[10]

contacting plane surfaces by machining grooves in one of the surfaces, preferably the one which is not subjected to the major cyclic stresses. This limits the extent of any particular contact and has been found to result in an improved fatigue behaviour. Figure 10.10 shows the effect of using grooved and plain fretting pads in contact with fatigue specimens of an aluminium alloy.[11]

It is possible that this factor is a side effect where surfaces are roughened

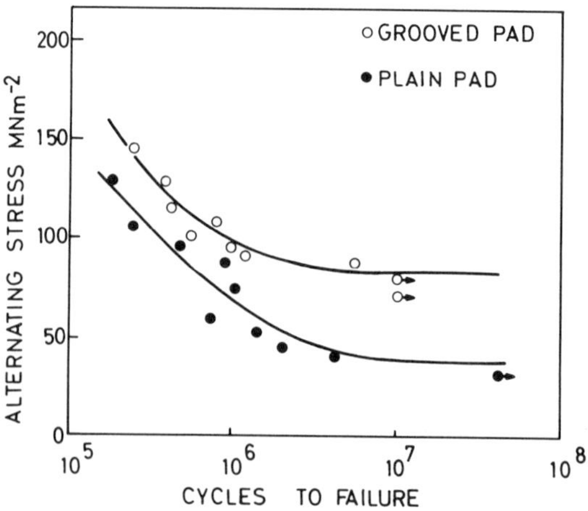

FIG. 10.10. Fatigue curves for an aluminium alloy fretted against plain and grooved fretting pads.

by shot peening and dimpling, although the major effect is usually attributed to the residual compressive stress in the surface as is described below.

These few examples indicate how the incidence of fretting fatigue failure can be greatly reduced by attention to the design of contacting components at the drawing board.

10.3 COATINGS

There is often some resistance to using applied coatings in fatigue applications since in many cases they reduce the normal fatigue strength. This is particularly the case with a coating that contains cracks, e.g. some hard chromium-plated coatings, or pores, e.g. sprayed molybdenum. Nevertheless where the coating produces a marked increase in the fretting fatigue strength its use can be justified.

10.3.1 Metal Coatings

Metal coatings have been applied to reduce fretting damage employing two widely differing concepts. The hard coating reduces individual areas of real contact and this in itself can be beneficial (see Section 10.2). However, the coefficient of friction between untreated metal surfaces is in the region 0·8–1·0 and thus the alternating shear stresses induced by the fretting action will be transmitted at these points of contact. The coating material must therefore have good fatigue strength under conditions of high strain fatigue. The greater the normal or clamping load the greater the depth to which these shear stresses are effective. The thickness of the coating will determine the magnitude of the stresses at the interface, which is often a plane of weakness. This has been found in the case of chemical-vapour-deposited titanium carbide, vacuum-deposited boron carbide, and electrodeposited cobalt–chromium carbide applied to titanium alloys.[12] In one case fretting occurred at the coating–substrate interface when the coating had debonded.

Soft metal coatings are applied with a view to exploiting their low shear strength, so that they act as a form of low friction bearing surface. They tend to be extruded from the contact region but this may not seriously detract from their function since only a very thin layer of metal is needed. Copper–nickel–indium coatings applied by the detonation gun method to a titanium alloy were extruded from the contact but nevertheless gave a fretting fatigue strength of 63 % of the normal fatigue strength, whereas the fretting fatigue strength of the untreated alloy was 48 % of the normal fatigue strength.

Syers[12] has investigated a wide range of metal coatings including bonded carbides, applied by such methods as electrodeposition, electro-less deposition, flame spraying, plasma spraying and detonation gun. His conclusion is that the hard coatings are generally most effective, e.g. cobalt bonded tungsten carbide applied by detonation gun. However, all the hard coatings reduce the unfretted fatigue strength and so a prior vapour-blasting or shot peening is necessary to counteract this. The intensity of the shot peening depends on the mode of application of the coating; for example, plasma spraying requires a more intense shot peening than the detonation gun method.

Electrodeposited coatings of chromium and nickel, as well as electro-less nickel, are significantly improved by a thermal diffusion process followed by shot peening of the treated surface.

Coating thicknesses in both the sprayed and deposited coatings were between 25 and 40 μm. With these treatments on Ti–6Al–4V the fretting fatigue strengths were in the range 50–60 % of the normal fatigue strength, whereas the fretting fatigue strength of the untreated material was 33 % of the fatigue strength of the untreated material.

The author has found that sprayed molybdenum coatings, although they reduce the fatigue strength of a carbon steel by 20 %, raise the fretting fatigue strength from 33 % to 72 % of the normal fatigue strength.[13] The hardness of the coating can be varied by variation in the oxygen content. In this case the coating hardness was 550 VHN compared with a substrate hardness of 210.

Many references will be found in the literature to the efficacy of hard metal coatings as a means of preventing fretting damage, but most of these have not been evaluated in fatigue. One such coating is the series of alloys known under a trade name but with a typical composition 50 % Ni, 32 % Mo, 15 % Cr and 3 % Sn. The structure is that of hard hexagonal Laves phases in a soft matrix, and they are said to have good anti-fretting and anti-galling properties. They have the advantage of being cobalt-free and therefore suitable for nuclear applications.[14,15]

The use of soft metal coatings is to be found much less frequently in the literature although one of the earliest anti-fretting treatments was the cadmium plating of automobile leaf-springs. Cadmium is still recommended together with other soft but low oxidising capacity metals such as silver and gold.[16] Such coatings have been applied to the critical blade-disc fixation in jet engines. Plasma-arc sprayed copper has been found effective on the footings of fan, compressor and low-pressure turbine blades.[17]

10.3.2 Non-metallic Coatings

The function of non-metallic coatings is to provide a surface which will not form the local welds which are a feature of the early stages of fretting between metal surfaces. The coating may take the form of a reaction product, e.g. phosphate coatings on steel or anodised coatings on aluminium and titanium alloys, or may be applied in the form of a monomer which is then polymerised by chemical action or heat treatment, e.g. certain epoxy resins. In some cases an insert of a thin sheet of nylon or polytetrafluoroethylene may be held mechanically between the surfaces, or better still, bonded to one or both surfaces. In the case of the hard coating the coefficient of friction is considerably lower than for metal-to-metal contact, i.e. in the region 0·2–0·3, but abrasion of an opposing metal surface may occur. However, abrasion without welding, although it may cause some fretting wear, has been shown to have little effect on fatigue strength.[18] Many of these coatings have a certain amount of porosity and can be saturated with an appropriate lubricant. The author has found that oil-in-water emulsions are effective with both phosphate and anodised coatings, especially where the stabilising polar molecule has a particular affinity for the coating.[19,20] The number of fretting cycles to initiate a propagating fatigue crack can be increased by a factor of five by this treatment. The effect of the polymer type coating depends on its elastic modulus. If this is low then the movement between the surfaces may be taken up in the coating itself. The more rigid materials can produce a form of fretting wear. In the case of steel surfaces this is due to the transfer of an oxide film to the polymer,[21,22] but there is no evidence that this has a detrimental effect on fatigue behaviour. In press-fitted steel unions, plain epoxy resin is said to increase the fretting fatigue strength by 30 % compared with the normal zinc paint previously applied.[23] Synthetic rubber coatings gave a 60 % increase in fretting fatigue strength on an 18Cr 9Ni stainless steel.[24]

Some investigators recommend the use of molybdenum disulphide with non-metallic coatings, either applied to the surface of the coating in the case of a phosphate coating,[25] or incorporated in the coating in the case of an epoxy resin.[26] The latter treatment restored the fretting fatigue strength of an aluminium alloy to the normal fatigue strength. Polytetrafluoroethylene coatings have also been filled with molybdenum disulphide or graphite and are effective when applied to only one side of the fretting contact. This treatment produced a 90 % increase in fatigue life on a low alloy steel but only 8 % on an aluminium alloy.[27]

Molybdenum disulphide when baked on to a surface could be regarded as a non-metallic coating in its own right, although it is usually regarded as a solid lubricant. Evidence of good wear properties in unidirectional sliding may not necessarily indicate that it is as efficient in preventing fretting, since its behaviour in reciprocating sliding leads to greater wear because of disruption of the overlapping alignment of the plate-like particles.[28]

Syers, in the work referred to previously,[12] has investigated non-metallic coatings on titanium. Anodising on two titanium alloys gave little or no improvement in fretting fatigue strength and the coating tended to spall easily. Covering the anodised coating with epoxy resin loaded with molybdenum disulphide produced no improvement. A polyimide resin loaded with boron nitride gave some improvement in the fretting fatigue strength of the alloy Ti–2·5Cu but none of the non-metallic coatings were as effective as the metal coatings described above.

10.3.3 Diffusion Coatings

Diffusion treatments can change the hardness and chemical nature of metal surfaces. Carburising and nitriding of steels produces an increase in hardness by increasing the concentration of the intermetallic compounds such as iron carbide which results in an increased pearlite content. An important feature of the process where fatigue is concerned is that it also induces a compressive stress in the surface. Vapour diffusion processes such as chromising and aluminising, change the chemical nature of the surface by substantial solid solution and thereby influence the corrosion and oxidation behaviour when applied to steel surfaces.

The author has shown that sulphidised coatings on steel are capable of raising the number of fretting cycles to initiate a propagating fatigue crack by a factor of five.[19] Subsequent successful practical applications of this coating to various industrial components, e.g. the fir-tree fixing of compressor blade roots, have reinforced the author's opinion that this coating is particularly effective on steels where the temperature of the treatment has no deleterious effect on the mechanical properties of the steel.

Diffusion coatings of boron, chromium and vanadium on steel, particularly after quenching and tempering at 200 °C, reduce the fretting damage, but these coatings have not been assessed in fatigue.[29] The vanadium coating reduced the damage to a very low value.

Diffusion treatments applied to titanium alloys were not particularly successful since they reduced the unfretted fatigue strength by factors of two to three.[12] Vapour blasting after treatment improved the situation somewhat but they were generally not as effective as other surface coatings.

10.3.4 Ion Plating and Ion Implantation

These two methods of surface treatment have come to the fore in recent years. Both depend on the acceleration by an applied voltage of ions from a target to the surface to be treated under vacuum conditions. In the case of ion plating, the accelerating voltage is relatively low and the ions do not have sufficient energy to penetrate the surface and consequently remain on the surface. The process is in this respect similar to vapour deposition. With ion implantation the ions are accelerated to much higher energies and penetrate the surface to depths of 0·25 μm. The surface is enriched by the implanted ions but a separate layer is not formed. It resembles the diffusion type of coating in that the chemical nature of the surface is changed, but it differs in two important respects in that the implanted ions cause considerable atomic disturbance and also that they remain mobile in the structure. It does, however, produce compressive stresses in the surface.

The beneficial effects of ion implantation in reducing unidirectional sliding wear are impressive[30,31] but its fretting resistance has not been assessed. Ion plated films of gold on steel surfaces are, however, much more resistant to fretting damage than sputtered or vacuum evaporated gold films (Fig. 10.11).[32]

As yet, few experimental data are available on the behaviour of these types of film in fretting fatigue. Some preliminary results on the alloy Ti–6Al–4V implanted with barium, caesium, strontium and ytterbium indicate that the fretting fatigue strength is increased from 33 % to 47 % of the normal fatigue strength.[33] This performance is comparable with that of

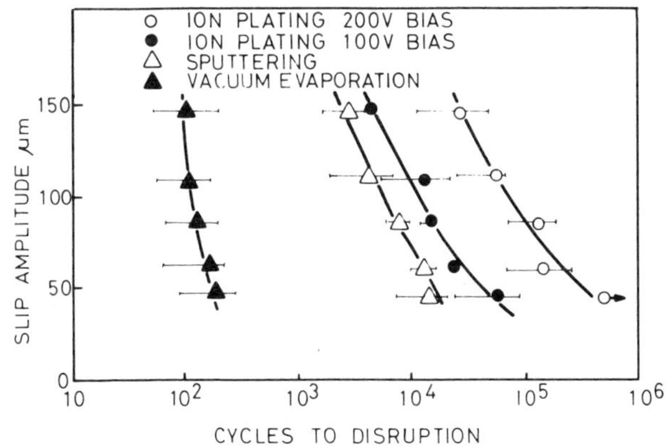

FIG. 10.11. Durability of gold films deposited on steel by various methods as a function of slip amplitude.[32]

the bonded carbide coatings described above. Diffusion treatment of the barium-implanted coatings at 750 °C had a deleterious effect.

10.4 LUBRICANTS

In certain circumstances fretting wear can be alleviated by increasing the access of liquid lubricants, e.g. in flexible couplings and spline connections, but in most of the instances where fretting fatigue is involved the assembly is of a permanent nature, e.g. a press- or shrink-fit or a bolted or riveted joint, and fluid lubrication would be impossible. The use of greases in such applications is also impracticable since the grease is likely to be squeezed out of the contact and adequate feeding arrangements would be difficult. The one exception here is in the construction of locked coil steel ropes where the grease can be incorporated into the centre of the rope and the surrounding conforming strands of the rope form a closed system.[1] The only lubrication system which can be recommended with any degree of general application is the solid lubricant film. Here, too, there may be problems in say a press-fit where the fitting of the components together may disrupt the film and render it ineffective.

The solid lubricant which has received most attention is molybdenum sulphide. In the earlier monograph[1] the author reported that molybdenum disulphide was initially effective in preventing fretting wear but eventually the coating was worn away and protection ceased. This observation has been confirmed by measurements in which the energy consumed per cycle of fretting is determined as a function of the total number of cycles.[34] The

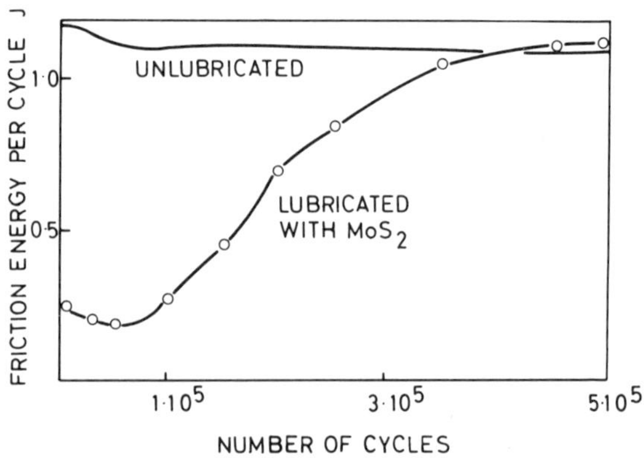

FIG. 10.12. Friction energy dissipated per cycle as a function of number of cycles for untreated and molybdenum disulphide coated steel surfaces.[34]

coated annular steel specimens were tested at an amplitude of 50 μm and a frequency of 20 Hz under a normal pressure of 20 MN m^{-2}. The results are shown in Fig. 10.12, which indicates that after 4×10^5 cycles the film is apparently worn through and the friction energy per cycle is the same for the untreated steel surfaces. This author recommends bonding the molybdenum disulphide to the surface with a resin as has been described in the section on non-metallic coatings. Although the assessment of molybdenum disulphide was made in fretting wear it can be assumed that if there is none or very little wear then the coating is likely to be effective in improving fretting fatigue performance. However, the short life of the coating does not recommend it for fatigue applications.

10.5 SURFACE COLD WORKING

This type of treatment includes shot peening, vapour blasting, bead blasting, surface rolling, and ballising of holes. Its effects are two-fold; it increases the hardness of the surface, the extent being dependent on the work-hardening capacity of the material, and it puts the surface layers in compression, both of which are advantageous where fretting fatigue is concerned.

The author has investigated the effect of shot peening on the fatigue strength of an austenitic stainless steel with the results shown in Fig. 10.13.[35]

The shot peening produced no effect on the plain fatigue behaviour of the steel, but the effect in fretting fatigue was most marked with the fatigue strength restored to the value without fretting. The surface hardness was increased to 400 VHN compared with the bulk hardness of 150 VHN. Shot peening of a mild steel, which has a lower work-hardening capacity than stainless steel, raised the fretting fatigue strength from 60 % to 80 % of the normal fatigue strength. In Fig. 10.13 it can be seen that specimens tested in fretting fatigue at stresses above the fatigue strength had longer lives than those tested in ordinary fatigue at the same stress levels. The reason for this unusual result is that the fretting still results in the initiation of cracks in the shot-peened surface but the compressive stress inhibits their growth. The depth to which they penetrate before ceasing to grow depends on the point below the surface where the compressive stress balances the alternating shear stress due to the fretting action. The effect of a number of surface cracks is to lessen the stress intensity at their tips.[36] The non-propagating cracks initiated in the fretted area have an influence on the initiation and propagation of cracks in the adjacent areas of the surface, and thereby

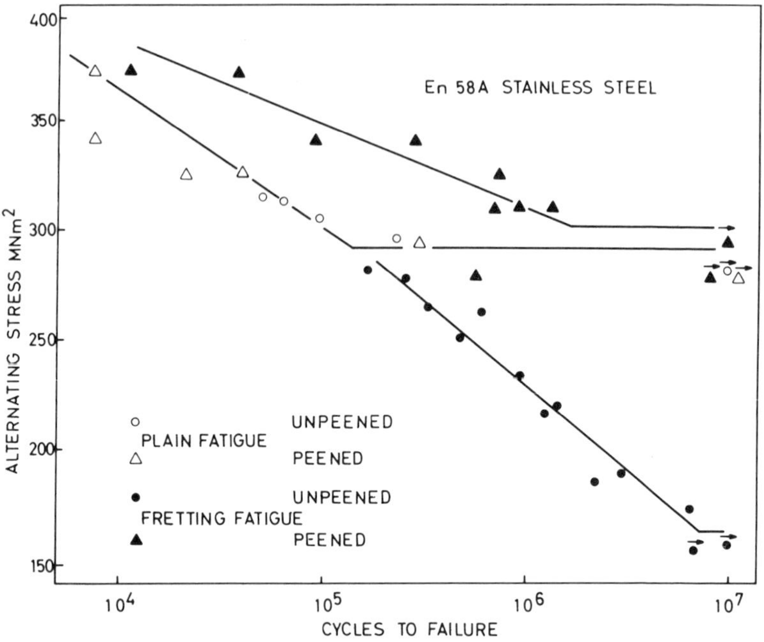

FIG. 10.13. Effect of shot peening on the fatigue behaviour in rotating–bending of an austenitic stainless steel (En 58A) with and without fretting.

reduce the area available for a propagating crack to develop and hence reduce the probability of such a crack being formed at a given stress level.

Shot peening has also proved beneficial on low alloy steels, where it produced surface residual stresses of 300–400 $MN\,m^{-2}$ to a depth of 120–150 μm. In this case also, the fretting fatigue strength was raised to the normal fatigue strength.[37]

Shot peening slightly reduced the fatigue strength of the alloy Ti–6Al–4V to 93% of the normal value but heavy shot peening reduced it to 57%.[12] However, with fretting the effects were similar, both treatments showing about 55% of the normal fatigue strength. A treatment described as 'vibrational shock' and presumably related to peening, is said to improve the fretting fatigue resistance by 70% when applied to titanium alloys.[38]

Another cold working treatment, dimpling, at two levels of intensity, was applied to an Al–4Cu–1Mg alloy. The ratio of the fatigue strength at 10^6 cycles of the undimpled specimens to the dimpled specimens was 0·94 for high pressure dimpling and 0·75 for low pressure.[39] In some recent work in rotating–bending on shot-peened specimens of a similar alloy the fretting fatigue strength was 34% *above* the normal fatigue strength whereas the untreated fretting fatigue strength was 55% of the normal fatigue strength. The residual surface stress in the shot-peened specimens measured by X-ray diffraction was 220 MN m^{-2}.[40]

Surface cold working can generally be recommended since it appears to have no adverse aspects. Shot peening can be used on its own or as an added treatment where coatings would otherwise impair the fatigue performance of the material, as has been seen in the section on metallic coatings. It must be remembered, however, that in an application where a temperature rise is possible, annealing may occur and the beneficial effects would be lost.

10.6 CHOICE OF MATERIALS

When selecting a material which has to operate under conditions of corrosion fatigue it is safe to say that a corrosion-resistant material will be a much more likely candidate than one which is not so. Is there a material which is fretting-resistant? The evidence is contradictory. Hard materials are resistant to fretting wear but experience shows that they may nevertheless suffer severe loss of strength in fretting fatigue. Fretting fatigue behaviour is very much related to notch sensitivity. Materials with low notch sensitivity, or the property more frequently referred to currently, high fracture toughness, are predominantly as-cast, annealed or tempered structures rather than work hardened or precipitation hardened. An annealed 70/30 brass is almost insensitive to fretting fatigue damage but its overall strength and fatigue strength are so low as to be of little application as a constructional material.

The susceptibility of a large number of materials, all tested in the author's laboratory in rotating–bending with the fretting surfaces of the same materials, i.e. like on like, is illustrated in Table 10.1.

Despite a wide range of UTS values and a related range of fatigue strengths, the fretting fatigue strengths are within a narrow band, the upper limit being between 150 and 160 MN m^{-2}. Strengthening methods which can be applied to increase fatigue strengths do not have a corresponding effect in fretting fatigue. It is for this reason that modifications of design and surface treatments are the only solutions available to the design engineer.

Where it is possible to choose dissimilar metals as members of a fretting contact it is possible to exercise some discretion. In an earlier work the

TABLE 10.1

FATIGUE STRENGTHS (f.s.) AND FRETTING FATIGUE STRENGTHS (f.f.s.) OF MATERIALS FRETTED AGAINST THEMSELVES

Material	*UTS* ($MN\,m^{-2}$)	*f.s. at* 10^7 *cycles* ($MN\,m^{-2}$)	*f.f.s. at* 10^7 *cycles* ($MN\,m^{-2}$)
0·2C steel	509	255	150
0·4C steel	772	266	87
0·7C steel (normalised)	940	216	154
0·7C steel (cold drawn)	1 360	525	139
18/9 austenitic stainless steel	598	295	160
70/30 brass (annealed)	328	89	80
70/30 brass (work hardened)	508	139	90
Al–4Cu–1Mg–1Mn	430	108	60
Al–4Mg–0·7Mn	314	100	73
Nickel-based alloy Inconel 718	1 316	550 ± 275	550 ± 120[a]
c.p. titanium (IMI 130)	675	201	123
Ti–2·5Cu	578	278	123
Ti–6Al–4V	1 095	540	154
Ti–4Al–4Mo–2Sn–0·5Si	1 125	634	139
Ti–6Al–5Zr–0·5Mo–0·25Si	1 054	432	154

[a] Tested in fluctuating tension.

author showed that soft aluminium (a clad alloy surface, VHN 31) required 23 times as many fretting cycles as Al–Zn–Mg alloy (VHN 182) to initiate a propagating fatigue crack in 0·16 steel (VHN 190).[1]

10.7 CONCLUSION

Remedial measures are available for counteracting fretting fatigue failures. Improved design is the method to be recommended wherever possible. However, where circumstances do not permit this, then surface treatments should be considered. Shot peening is particularly efficient for room or moderate temperature operation. Coatings of various kinds have to be treated with some circumspection because of their propensity to reduce the fatigue strength of the material, but treatments which induce a residual stress in the surface, such as nitriding and sulphidising can be recommended. Alternatively shot peening can be applied before coating to restore the original fatigue properties.

REFERENCES

1. Waterhouse, R. B. *Fretting corrosion*, Pergamon, Oxford (1972).
2. *Control of Fretting Fatigue*, National Materials Advisory Board, Report No. 333, Dept. of Defense, Washington D.C. (1977).
3. Gordelier, S. C. and Chivers, T. C. A literature review of palliatives for fretting fatigue. *Wear*, **56**, 177–90 (1979).
4. White, D. J. *Proc. Instn. Mech. Engrs.*, **185**, 709–16 (1970–71).
5. Buch, A. *Wear*, **43**, 9–16 (1977).
6. Gassner, E. *Aluminium*, **39**, 582–4 (1963).
7. Sandifer, J. P. *Wear*, **26**, 405–12 (1973).
8. Hamel, D. R., Korbacher, G. K. and Smith, D. M. *Trans. ASME, J. Basic Engrng. Series D*, **39**, 649–56 (1971).
9. Jergeus, H. A. Euromech, Colloquium No. 110, *Contact Problems and Load Transfer in Mechanical Assemblages*, 109–13 Linköping Institute of Technology (1978).
10. Morton, P. G., Goodman, P. J. and Kawecki, Z. M. *Proc. Instn. Mech. Engrs.* **184** Pt. 3B 66–74 (1969–70).
11. Bramhall, R. *Studies in fretting fatigue*, Ph.D. Thesis, Oxford University, November 1973.
12. Syers, G. The Protection of Certain Titanium Alloys against Fretting-Fatigue. Rolls-Royce Laboratory Research Report 10245/FS/78, p. 118.
13. Taylor, D. E. and Waterhouse, R. B. *Wear*, **20**, 401–7 (1972).
14. Anon. *Tribology International*, **10**, 254 (1977).
15. Schmidt, R. D. and Ferris, D. P. *Wear*, **32**, 279–89 (1975).
16. Moreau, M. and Gostoli, M. *Matér. Tech.*, **64**, 385–92 (1976).
17. Hermanek, F. J. Jr., *Metal Progress*, **97**, 104–6 (1970).
18. Fenner, A. J. and Field, J. E. *Rev. Mét.*, **55**, 475–85 (1958).
19. Waterhouse, R. B. and Allery, M. *Wear*, **8**, 112–20 (1965).
20. Waterhouse, R. B. *Tribology*. **3**, 158–62 (1970).
21. Higham, P. A., Stott, F. H. and Bethune, B. *Corros. Sci.*, **18**, 3–13 (1978).
22. Higham, P. A., Stott, F. H. and Bethune, B. *Wear*, **47**, 71–80 (1978).
23. Nishioka, K. and Komatsu, H. *J. Soc. Mat. Sci. Japan*, **19**, 988–93 (1970).
24. Ivanova, V. S. and Veitsman, M. G. *Fatigue of metals and alloys*, 103–8, NAUK, Moscow (1971).
25. Bollani, G., Castagna, M. and Ferrero, P. *La Metallurgia Italiana*, Nos. 7/8 366–72 (1976).
26. Harris, W. J. D. Mat. Report No. 174, July 1971.
27. A. Ya. Alyab'ev *et al.*, *Sov. Mat. Sci.*, **13**, 177–80 (1977).
28. Holinski, R. and Gänsheimer, J. *Wear*, **19**, 329–42 (1972).
29. Demko, A. A. *et al.*, *Sov. Mat. Sci.*, **8**, 9–11 (1972).
30. Hartley, N. E. W. *Thin Solid Films*, **64**, 177–90 (1979).
31. Dearnaley, G. *Proc. Mat. Res. Soc.*, Cambridge, Mass, November 1979.
32. Ohmae, N. *J. Vac. Sci. Technol.*, **13**, 82–7 (1976).
33. Syers, G. A Preliminary Investigation into the Effects of Ion Implantation on the Fretting Fatigue Behaviour of Titanium Alloy Ti–6Al–4V. Rolls-Royce Laboratory Research Report 23941/P1/79.

34. Müller, K. *Tribology International*, **8**, 57–64 (1975).
35. Waterhouse, R. B. and Saunders, D. A. *Wear*, **53**, 381–6 (1979).
36. Waterhouse, R. B. *Corrosion*, Ed. L. L. Shreir, Newnes–Butterworth, London, 2nd Edition 1976, 8:97–113.
37. A. Ya. Alyab'ev *et al.*, *Technol. Organ. Proizvod. Naucho-Proizvod.* Sb. (6) 11–12 (1976).
38. Veitsman, M. G. and Vainshtein, V. G. *Vestn. Mashinostr.*, Sept. 1975, 73–5.
39. Lohse, A. and Shewchuk, J. Proc. Society for Experimental Stress Analysis Meeting, SESA San Francisco 20–25 May 1979.
40. Leadbeater, G., Noble, B. and Waterhouse, R. B. (to be published).

The work referred to in references 12 and 33 is by kind permission of Rolls-Royce Ltd (Aero Division) and the Ministry of Defence (Royal Aircraft Establishment).

Index